Sadguru Model of Rural Development
Elevates Food Security and Ease Poverty

The Author

Govindasamy Agoramoorthy is Distinguished Research Professor at the College of Environment and Health Sciences in Tajen University, Taiwan. His research ranges from environmental science to sustainable development, and he has carried out field research related to natural resource management in Asia, Africa, and South America over the last three decades. He serves in the editorial board of several international peer-reviewed journals including the Journal for Nature Conservation (Elsevier), Frontiers in Earth Science (Nature Publishing Group) and Journal of Environmental Biology (Triveni). He served as Visiting Scientist at the Smithsonian Institution in Washington DC (USA) between 1989 and 1993. He is currently Tata Visiting Chair at Sadguru Foundation, Gujarat, India. Professor Agoramoorthy has authored 25 books, 80 book chapters, and 300 scientific articles in peer-reviewed journals with impact factor.

Sadguru Model of Rural Development

Elevates Food Security and Ease Poverty

Govindasamy Agoramoorthy

Distinguished Research Professor
Tajen University, Yanpu, Pingtung, Taiwan

Tata Visiting Chair
NM Sadguru Water and Development Foundation
Dahod, Gujarat, India

2016

Daya Publishing House®
A Division of
Astral International Pvt. Ltd.
New Delhi – 110 002

Cataloging in Publication Data--DK
Courtesy: D.K. Agencies (P) Ltd. <docinfo@dkagencies.com>

Agoramoorthy, Govindasamy, 1957- author.
Sadguru model of rural development : elevates food security and ease poverty / Govindasamy Agoramoorthy.
pages cm
Includes index.

ISBN 978-93-5130-960-4 (International Edition)

1. Rural development--India--Rajasthan. 2. Rural development--India--Gujarat. 3. Rural development--India--Madhya Pradesh. 4. Agriculture--Economic aspects--India--Rajasthan. 5. Agriculture--Economic aspects--India--Gujarat. 6. Agriculture--Economic aspects--India--Madhya Pradesh. 7. Food security--India--Rajasthan. 8. Food security--India--Gujarat. 9. Food security--India--Madhya Pradesh. 10. Poverty--India--Rajasthan--Prevention. 11. Poverty--India--Gujarat--Prevention. 12. Poverty--India--Madhya Pradesh--Prevention. 13. Sadguru (Organization : India). I. Title. II. Title: Elevates food security and ease poverty.

HN690.Z9C6 2016   DDC 307.14120954   23

Published by   :   **Daya Publishing House**
                              A Division of
                              Astral International Pvt. Ltd.
                              – ISO 9001:2008 Certified Company –
                              4760-61/23, Ansari Road, Darya Ganj
                              New Delhi-110 002
                              Ph. 011-43549197, 23278134
                              E-mail: info@astralint.com
                              Website: www.astralint.com

Laser Typesetting   :   **Classic Computer Services,** Delhi - 110 035

Printed at   :   **Thomson Press India Limited**

— Dedicated to —

India's Legendary Rural Development Stalwarts,
Shri Harnath Jagawat and
Smt Sharmishta Jagawat

For sharing their immense knowledge and compassion

The support of Sir Dorabji Tata Trust (Mumbai) to conduct research at Sadguru Foundation through the Tata-Sadguru Visiting Chair Status awarded to the author is greatly appreciated!

Acknowledgments

"Gratitude is the fairest blossom which springs from the soul"

— Henry Ward Beecher

I sincerely thank Shri Harnath Jagawat and Smt. Sharmishta Jagawat for their warm hospitality and kind support over the years during my numerous visits to Sadguru Foundation. I am fortunate to have been guided by these two legendary rural development stalwarts. Their vision, experience and willingness to advise me have enormously benefited the production of several scientific papers including this book.

The contributions of my friends Kanhaiya Choudhary, Sunita Chaudhary, Hitesh Shah, Ramesh Patel and Shodhan Shah for sharing their experiences on irrigation, check dams, vegetable cultivation and financial management are greatly appreciated. A few hundred villagers and a few dozen employees associated with Sadguru to whom I had interacted in Gujarat, Rajasthan and Madhya Pradesh states, answered all my curious queries. They are too many to name individually, and I thank them all for their kind cooperation. On each field trip, I had the unique fortune to encounter an array of good-hearted people who have assisted me in many ways, from offering tea to sharing their knowledge.

Several tribal youngsters who work tirelessly at Sadguru's kitchen in the Chosala campus, especially the executive chef, Nagin took special care

Author and friends from
Sadguru Foundation, India

of me by preparing delicious and less-spicy health food on a timely fashion. I thank those kitchen foot-solders for their hard work. Senior Sadguru staff members endured my investigations into their professional activities with grace, patience and extraordinary openness, and I am grateful to all of them. Finally, I thank the financial support of Sir Dorabji Tata Trust (Mumbai) for awarding me the Tata Visiting Chair status to carry out research at Sadguru Foundation for many years.

Govindasamy Agoramoorthy

Preface

"Civilization as it is known today could not have evolved, nor can it survive, without an adequate food supply"

— *Norman Borlaug*

India's workforce mainly relies on farming despite the agriculture contribution to GDP has diminished from 38 per cent in 1975 to 13.7 per cent in 2013. India requires 210 million tons of grain to feed its people now. But, it produces about 200 million tons each year. As a result, the flour factories across India are now concerned that there is not enough wheat in the market to meet the demand, which led them to place import orders of half million tons of wheat from Australia in March 2015. The cultivable land has remained stagnant at 120 million ha relying mainly on monsoon water. So, the country needs innovative thinking to diversify agriculture.

One of the major factors for the agro-crisis is the population pressure that has outstripped the land supply. Faster growth of population goes with slower rise in per capita output. Due to the rapid expansion of industrial development, the pressure on land is being further aggravated. The other reason attributes to inefficient distribution, marketing and financial institutions. As agriculture falls under the domain of the informal sector, there is a shortage of efficient institutions. The loans for farmers are sometimes hard to come by as they lack assets for collateral. Furthermore, the current rate of inflation may lead to price hikes that could further complicate life for farmers. The distribution systems fail as agro-products often doesn't reach targeted population.

Other problems contribute to the agriculture crisis include; mining and submersion of land due to construction of large dams, lack of underground water to grow water-demanding crops, government encouragement to cultivate medicinal/aromatic plants reducing area of staple food crops, deficiency in Research and Development to increase productivity, improper chemical fertilizer recommendations for crops based on old soil surveys, lack of crop insurance for marginalized farmers, politicians with poor agro-tech vision, bureaucratic stumbling blocks for farmers' growth, technological gap, poor strategic planning, and lack of farmer suicide prevention measures. Under these circumstances, is there any change to achieve sustainable agriculture in India?

If India needs to improve farming, it must embark on new initiatives to boost resource development by incorporating people, land and water with government, non-government and corporate support. On the whole, the condition of India's current farming condition is weak. The sector is pressurized by the domestic and global economy. Agriculture finds it hard to compete in global markets with other products that are highly subsidized. Hence more funds need to be devoted for research to improve productivity using the genetic revolution. The size of agricultural markets needs to be increased; it can be facilitated by the extension and improvement of transport. Pumping money for rural areas should go with targeted policies to bring favorable outcomes to enhance agriculture production in villages and eradication of poverty at the grassroots.

An important constituent of the strategy to revive India's farming performance must be to increase the level of public investment in agriculture research, technology and rural infrastructure. Increase in agriculture investment in turn requires rationalization and restructuring of subsidies on food and inputs, including power, irrigation and fertilizers. India does not produce enough electricity to sustain farm productivity.

The participatory management and selective privatization may be essential to improve the delivery of inputs to agriculture. Until recently, farmers were forced by law to sell their produce at local markets aimed at protecting poor farmers from exploitation. But, it's now controlled by cartels of traders, trivial bureaucrats, and loan sharks. Produce from villages are sent through middlemen on a slow and often hot journey to retail outfits hundreds of miles away and in between 40 per cent of it rot before reaching the final destination. Due to few refrigerated packing centers with no distribution network and inefficient fleet of trucks, India can no longer sustain large-scale vegetable production and transport.

With the majority of India's population relying on agriculture, improvements in efficiency and markets have the potential to benefit people. The initiatives by major corporations will bring advantages of scale that have largely been missing in a nation where the average land holding is only 2.5 acres and about 60 per cent of farming output is consumed by the families. Industries are aware of the fact that if business lead to consolidation or farmers' being displaced from their traditional land, it will become politically-sensational at a time when crop failures and bankruptcy lead to an average of 15,000 farmer suicides per year.

By 2050, the global human population pressure will demand resources at double the rate at which the Earth can generate. Five major factors that determine the extent of global overshoot or demand on bio-capacity includes population, consumption of goods/services per person, resource use intensity, bio-productive areas, and bio-productivity per hectare. India has to make plans with persistence to eliminate the overshoot of the above five factors from now onwards. One crucial aspect to eliminate overshoot of bio-productive areas is transforming India's vast drylands.

I have visited the tribal areas in Gujarat, Rajasthan and Madhya Pradesh states several times, and shocked to see villages without electricity, infrastructure, sanitation and other basic needs of humanity. These villages remain less progressive since they are located in unproductive dryland where the tribal communities live. Government officials seldom venture out to explore villages due to logistical nightmare in the remote wilderness. Drylands receive only 4-days of rain per year and that too after 3-years of severe drought. Thus saving every rain drop is crucial to sustain humans and other life forms in this harsh wonderland. Can water actually eliminate poverty in tribal areas?

Headed by India's renowned rural development social work couple, Harnath Jagawat and Sharmishta Jagawat, and armed with a team of dedicated dam-building and irrigation engineers, the Sadguru Foundation started to build series of cost-effective lift irrigation systems and check dams in rivers across the tribal areas with a sole purpose to hold water for irrigation. Till March 2015, a total of 376 such check dams and 401 lift irrigation systems in tribal drylands were constructed, which in turn converted over 100,000 acres of wasteland to productive agricultural land through community irrigation cooperatives. This massive effort supported by other activities ultimately benefited nearly 2 million people in tribal villages of India. These families used to live in absolute poverty prior to the intervention. But now each family saves an average of 10,000 Rs. per year and most people are satisfied with this economically enhanced life.

Finding water in the drylands of India is a luxury, but the Jagawats have succeeded in not only harvesting water using check dams on rivers but also lifting water from low-lying sources (tanks, ponds, and rivers) to upland farms. They have dutifully followed the national father, Mahatma Gandhi's vision of village business and rigorously tested his economic principle in the field using the power of water to sustain and reward life. Traditionally, people in tribal areas grow one seasonal red-fed crop. Droughts plague the drylands every few years and this harsh reality triggers people to migrate to towns and cities in search of work. After check dams and lifts irrigation cooperatives were established, migration stopped, crop productivity increased and poverty decreased. Cooperatives manage infrastructures and farmers pay a modest sum for the use of water to irrigate their fields.

The village-based sustainable development activities of Sadguru foundation involve schemes associated linked to check dam, lift irrigation, renewable energy, horticulture, floriculture, dairy farming and other income generating work directly and indirectly contributing to food security, poverty reduction and climate change mitigation at local and regional levels.

As showed in this book, transforming the unproductive drylands across India to achieve a contemporary green revolution to elevate food security and ease poverty is indeed possible in rural India with the model of 'Sadguru Model of Rural Development' portrayed in this book.

Govindasamy Agoramoorthy, Ph.D.

Distinguished Research Professor
Tajen University, Yanpu, Pingtung, Taiwan

Tata Visiting Chair
NM Sadguru Water and Development Foundation
Dahod, Gujarat, India

Email: profagoramoorthy@yahoo.com

Contents

Chapter 1

Partnership for Poverty Reduction

"The ultimate goal of farming is not the growing of crops, but the cultivation and perfection of human beings"

— *Masanobu Fukuoka*

Introduction

The World Summit on Sustainable Development has identified four major sectors such as the government, corporate, NGO and community as key players for poverty and hunger reduction targets in developing countries (Warner and Sullivan, 2004). If a successful partnership involving these sectors can be built on the idea of each partner doing what they do best, the combined approach may reduce poverty faster. Although such an alliance may enable communities to take charge of their own developmental needs to improve livelihoods, distrust among various partners continue to exist (Heap, 2000). Therefore, the challenges of poverty reduction are complex that not one sector alone can address this issue on its own (Backstrand, 2006).

The major downside of India's agriculture strategy is the neglect of dryland catchment areas about 1,500 km long and 500 km wide stretching from Dungarpur in the west to Dumka in the east that harbors 67 million people belong to the indigenous communities. They are the poorest since they have less access to public services (Agoramoorthy, 2008). Foremost among the challenges they face are low rainfall, frequent droughts, poor soil condition, and unreliable water supply. If India needs to reduce rural poverty, it must reorient development by integrating indigenous people, land, and water and work in partnership with government, non-government organizations (NGOs) and corporations (Agoramoorthy, 2009).

Forming partnership to reduce poverty is not easy since it demands trust, transparency and flexibility (Brinkerhoff 2002). Besides, there are challenges involving bureaucracy, trust building and capacity-building since they play a major role in maximizing partners' interactions (Lister 2000; Brinkerhoff, 2002). If a NGO establishes goodwill and reputation even with a single agency, it has the potential to attract other partners. This chapter presents partnership involving communities, government agencies, NGOs and corporations by drawing on data collected from a case study of a non-profit agency in Gujarat, India. This chapter also presents challenges and opportunities for the NGO sector involved in water resource development to enhance a successful partnership to reduce poverty in India's drylands.

Background

From the early 1960s to late 1970s, 5 major groups of NGOs became active in India (Sen, 1999). The first consisted of educated middle-class Indians dissatisfied with the government while the second involved in the modernization. The third composed of the Sarvodaya Movement, and the fourth involved the Naxalite or Maoist-Leninist armed movement. Between 1980 and 2010, 6,000 people died due to violence involving the Naxalites. The fifth comprised of academics influenced by the Marxist ideology (Mehra, 2000). Politicians, retired bureaucrats, industrialists and young professionals also formed NGOs in the 1980s and 1990s. At the same time, the relationship among the government and NGO sector took a turn with issue-based campaigns involving large dams, environmental destruction, housing shortages and women's rights. A notable example was the anti-dam campaign in 1986 initiated by Save Narmada Agitation. Activists launched campaigns and legal battle for the rights of millions displaced by large dams (Kothari, 1986). Due to such provocative campaigns, the government and corporate sectors started to perceive NGOs as stumbling-block for development (Zaidi 1999; Anand and Anand 2007). Despite this pessimism, the 8th Five Year Plan (1992-1997) of the Indian government asked for more participation of NGOs in poverty reduction schemes (Government of India 1992a, b).

Importance of Partnership

In January 1992, India's Prime Minister (N Rao) said that he was prepared to assign development projects to NGOs and told them to come forward if they faced bureaucratic problems (Singh, 1992). Similarly, the Krishnaswamy Commission reviewed government-NGO partnership in 1988 and recommended to simplify procedures so that NGOs could get grants faster (AVARD, 1991). Such support from the government subsequently led to the outburst of NGOs. According to the Central Statistical Organization (2010), 3.3 million NGOs exist across India of which 45 per cent came into effect after 2001. It is widely recognized that good governance relies on social capital within communities by involving people in public processes (Evans, 1996; World Bank 1997). It is also known that NGOs play a key role to improve social capital for the society. They can contribute to state-funded public services by providing tactical support making services effective (Young, 2000). Partnership is a major topic of discussion in numerous conferences and notable donors namely the World Bank and UNDP have sought to improve relations between governments, NGOs and corporations to create public services for poverty solutions (Coston, 1998). The DAC 21st strategy, the UN Common Country Framework, and the World Bank's Comprehensive Development Framework and Poverty Reduction Strategy have highlighted partnership as a means to reduce poverty (OECD 2001; World Bank, 2001).

The questions, however is, what is the meaning of partnership in the context of poverty? According to Cohen and Uphoff (1977), partnership is initiated from below, voluntary, organized, empowering and continuous in scope. Although the importance, relevance, and ways of operation may vary among partners, the bottom-line is to achieve poverty reduction. Ashman (2001) suggests partnership involves deciding together and acting together with collaborative accountability where partners recognize and appreciates the contributions of all. Mutuality refers to a form of symbiosis, a biological term, in which two or more partners form a relationship to produce equal and enduring benefits (Austin, 2000). According to Warner (2002), partnerships for development are no different in principle from conventional business-to-business strategic alliances. Unlike contractual relationships, partnerships for rural development between business, government and NGO seek not to shift responsibility and risk from one party to another, but to share risks, resources and talents (Warner, 2002). While assessing the World Bank's Business Partners for Development, Price Waterhouse Cooper (2002) concluded that tri-sector partnership can deliver many of the same benefits currently attributed to conventional business alliances involving knowledge, specialist capabilities/resources and sharing of the risks/ costs of new ventures. This conclusion has reflected in a recent partnership study on India's health sector that states effective partnership among the public, private and NGO sectors can increase tuberculosis case notifications while maintaining acceptable treatment outcomes therefore partnership holds potential to manage contagious diseases (Dewan *et al.,* 2006). It has social workers in its employ that use community organization method in partnering with people in their community for development.

Although the Indian government has been showing interest to work with NGOs since the 1990s, not all in the sector trust the government (Rajasekhar and Biradar, 2004). It is perhaps due to the existence of two schools of thoughts. The opponents argue that the core ethics of voluntarism is undermined by government collaboration therefore NGOs should not waste time and resources to gratify politicians and bureaucrats (Jain 1989; Boris *et al.,* 1999; Ackerman 2008). On the other hand, the proponents of partnership argue that the government spends huge amounts of public money for development and some portion of it gets wasted/misused therefore it's necessary for NGOs to engage in partnership (Jain 1989; Warner and Sullivan 2004; Jagawa, 2005). But, would it be possible for NGOs to build up better working relationship involving the government, corporate, NGO and community sectors to reduce rural poverty? This chapter has addressed this question by presenting a case study on how a water-based NGO in Gujarat state of India managed to partner with government agencies, corporations, NGOs and communities to reduce poverty.

Methodology

The NM Sadguru Water and Development Foundation also known as Sadguru Foundation is a nonprofit agency based in Chosala village of Dahod District, Gujarat State, India. It has been involved in rural development since 1974 (Figure 1). The present case study was conducted in the drought-prone drylands spreading across the states of Gujarat, Rajasthan and Madhya

PROGRAMME AREA - STATES OF SADGURU FOUNDATION

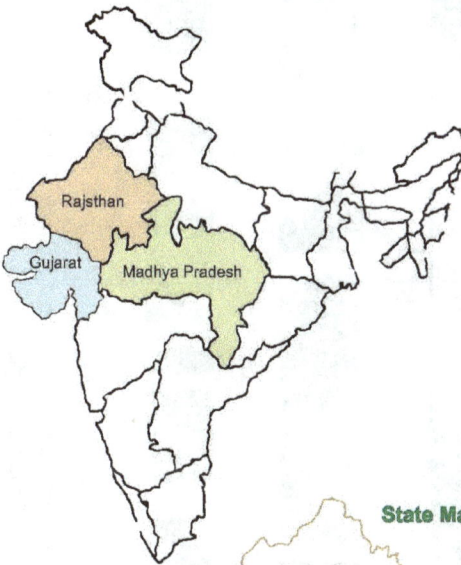

MISSION

SADGURU endeavors to develop and expand environmentally, technically and socially sound natural resource interventions leading to empowerment of rural community including women to ensure equitable and sustainable development and poverty reduction.

VISION

Empowerment of tribal and rural communities with natural resources restored, developed and expanded in the selected project areas.

Figure 1: Study Area and the Organization of Sadguru Foundation in India.

ORGANOGRAM OF ORGANIZATION

```
BOARD OF TRUSTEES
        |
ADVISORY COUNCIL
        |
DIRECTORS GENERAL
        |
    +---+---+
    |       |
CEO-CUM-DIRECTOR        DIRECTOR
OPERATION               FINANCE
```

- General Administration
- Finance Department
- Department of Water Resources
  - Check Dams
  - Lift Irrigation
- Department of Micro Watershed
  - SHGs Micro Finance
  - Agriculture Extension
- Department of Environment, Forestry, Horticulture etc.
  - Rural Energy
  - Sahaj
- Community Support Services & Other Programmes
- Department of Village Institutions & Co-operatives
- Department of Monitoring & Research
- Field Project Officers
  - Dhanpur
  - Danod
  - Jhalod
  - Limkheda
  - Banswara Rajasthan
  - Jhalawar Chaumahla Rajasthan
    - Shilkuva
    - Suwasara* Pidawa
- Training Institute

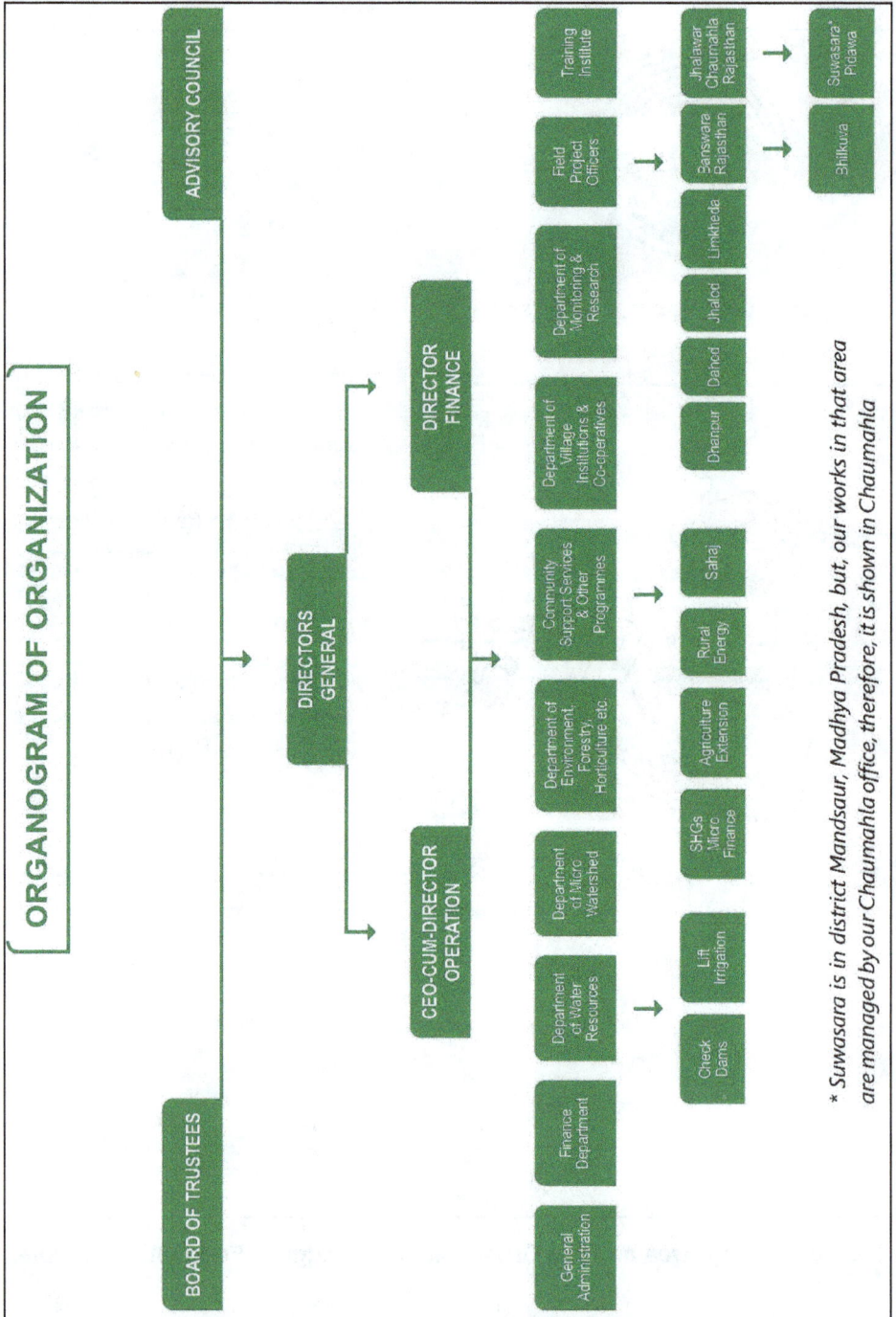

* Suwasara is in district Mandsaur, Madhya Pradesh, but, our works in that area are managed by our Chaumahla office, therefore, it is shown in Chaumahla

Pradesh in western India. The district of Dahod (3,642 km²) has a population of 1,636,433 and is the poorest in Gujarat with 72 per cent of the Bhil tribe (Government of India, 2011). It receives 860 mm of annual average rainfall. The Jhalawar District (6928 km²) in Rajasthan is one among India's poorest supporting a population of 1,180,342. Most of the villages do not have public services with the exception of few schools. The Jhabua District in Madhya Pradesh (6,782 km²) has a population of 1,396,677 (85 per cent indigenous of which 47 per cent suffer from extreme poverty; Government of India, 2011).

Check dams located in Gujarat, Rajasthan and Madhya Pradesh were visited between July 2006 and December 2008 to record data for this case study. Data on rivers, check dams, water storage, construction cost, household structure, job opportunities and irrigated area increase for 305 check dams constructed from 1990 to 2007 were pooled from Sadguru's archives. In addition, data on the migration patterns of farmers from villages to towns in search of work during non-farming season were collected by interviewing leaders of community cooperatives in 12 villages of Dahod district (Gujarat) and also by pooling data from their records (Mikkelsen, 1995). Data on funding that Sadguru received from donors were pooled from its archives. All statistical analyses were done using Statistical Analysis System software and all mean values are presented as ± 1 standard deviation (SAS Institute 2000). A linear regression analysis was used to estimate the cost (US$) according to the length of check dams. The student *t*-test was used to test the difference of men, women, total men/women and days of migration from 12 selected villages before and after the initiation of water resource development activities (SAS Institute, 2000).

Check Dams

India's drylands on average receive rainfall from 400 to 1000 mm annually and if it is harvested, it could support farming. Hence Sadguru embarked on water resource management work to foster socioeconomic development in villages. This hydro-centric approach included water harvesting through community check dams, pumping of water from low-laying rivers to higher-up farms via lift irrigation, digging/recharging of wells, developing micro-watershed schemes, and establishing drinking water supply. From 1976 to 2010, a total of 342 check dams and 341 lift irrigation systems were constructed, which in turn converting 310,225 acres of unproductive drylands to productive farmlands.

Check dams are small barriers using stones, cement and concrete built across rivers/streams for harvesting rainwater. They retain excess flow during rains in small catchment areas. Mechanized pumps lift water from low-lying dams to high-up farms called lift-irrigation system. According to the International Commission on Large Dams, any dam higher than 15m is considered "large dam," and when it exceeds 150m, it is considered major dam (Rangachari *et al.*, 2000). Check dams are smaller than 15 m and they are eco-friendly. Unlike large/major dams, check dams neither displace people nor submerge lands.

Sadguru spent US$12 million to construct 305 check dams between 1990 and 2007, which in turn saved 1,657.17 million ft^3 of rain water irrigating 47,700 acres of drylands. The average number of beneficiary was 47.49±174.26 households (range 2–3,000; 6 members on average each household) and these families expanded on average 156.48 ± 433.37 acres of irrigated area (range 3–7,000 acres). In 2007, the Chief Minister of Rajasthan inaugurated the Baneshwardham check dam (length 367 m; cost US$ 1.18 million in 2006 with a storage capacity of 350 million ft^3 with a potential to irrigate 7,000

Figure 2: The Chief Minister of Rajasthan State (Smt. Vasundhara Raje) is a Strong Supporter of Check Dams and she Inaugurated India's Largest Check Dam in 2007 (Photo courtesy Sadguru Foundation).

acres benefiting 18,000 farmers (Figure 2). Upstream from this check dam lays a large dam (Mahi-Bajaj Sagar costing US$ 300 million) with a potential to irrigate 154,000 acres, which is 22 times more than the irrigation area of the Baneshwardham check dam. If 22 more check dams were to be built in a series, it would cost US$ 24.2 million, with the potential to irrigate the same area of Mahi-Bajaj Sagar dam. If check dams are built in large numbers both upstream and downstream of major rivers, they will have similar potential as do large dams to expand irrigation. But, it would cost low and with less negative environmental and social impacts.

Jobs, Community Empowerment and Migration Reduction

Mobilizing people to establish village institutions is an important constituent of Sadguru foundation. By working with the community, it was able to create several village-level cooperatives to manage water-based agricultural projects. All cooperatives are legally registered under the Cooperative Societies Act. A total of 2,577 village institutions such as lift irrigation cooperative, check dam management group, joint forest management group, women self-help groups, farmers' group, watershed association, drinking water committee, dairy cooperative and horticulture group was created. These institutions have a total of 89,732 members and some became large federations by combining smaller cooperatives and positioned better to seek technical/financial assistance and to liaise with state and national-level donors. Over 1000 federations have been active with a membership of 500,000 farmers. These cooperatives, small and large, manage their own savings, raise funds from donors, give micro-credit to members, and manage water resource infrastructures. The creation of village cooperatives is fundamental not only to manage water infrastructures, but also to improve livelihoods of communities. Similarly, diverse social network of civic associations are known to confront poverty, resolve social disputes and provide opportunities for community development (Midgleya and Livermore 1998; Varshney 2000).

The total employment generated by Sadguru involving the construction of water harvesting structures from 1985 to 2010 was 430,719 person days with 62.7 per cent involvement of women. Local farmers usually relied on migration to nearby towns and cities in search of labour jobs since they were depending on rain-fed agriculture. After check dams were constructed, farmers had access to irrigation water leading to self-sufficiency in food production in villages that also eased migration. The average number of people migrating before in villages was 74.58 (± 40.36, range 18-143, n=12) with 178.83 average days (± 22.44, range 137-219, n=12; Figure 3). Afterwards, the number of people migrating to towns and cities reduced to 25.58 (± 13.30, range 6-48, n=12) with 57.5 average days (± 30.26, range 9-111, n=12). The reduction in migration showed gender differences as well, and for example, migration of men reduced by 61 per cent, from an average of 45.75 (±18.85, n=12) to 17.83 (22.44, n=12), while women migration reduced even higher by 73 per cent, from an average of 28.83 (±23.97, n=12) to 7.75 (7.03, n=12; Figure 3). This indicates that water resource development projects have great potential to improve irrigation, minimize migration and maximize livelihood opportunities for people who inhabit the drylands of India.

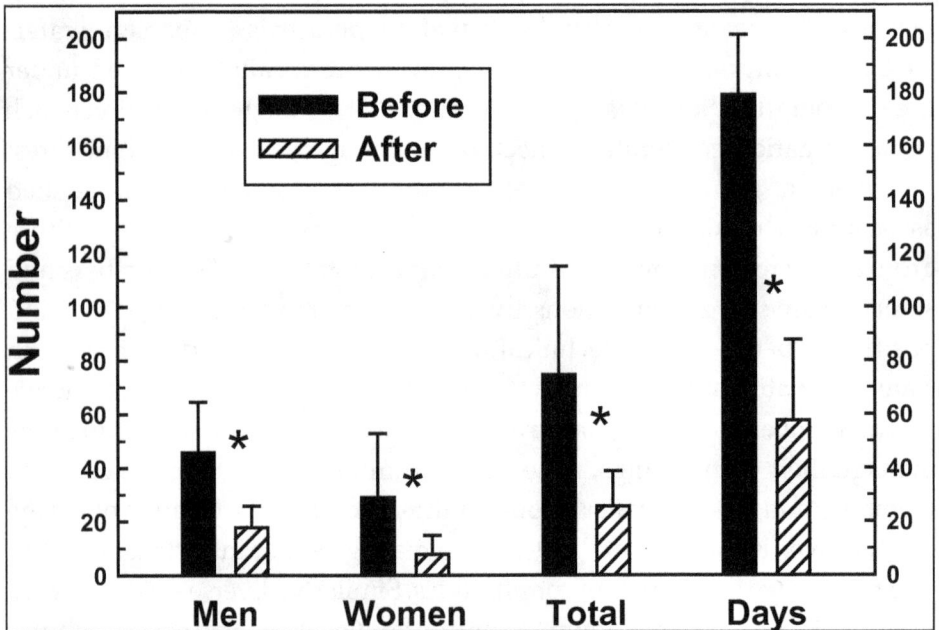

Figure 3: Average Number (± SD, n=12) of Men and Women Migration, and Total Days of Migration from 12 Villages before and after Water Resource Development Work by Sadguru Foundation (*: Pair-t test, p<0.05).

Funding and Accountability

The average funding received by Sadguru between 1997 and 2008 was US$ 3.19 million (± 1.25, n=12). It grew over the years from US$ 1.63 million in 1997 to US$ 2.89 million in 2001 contributing to an overall increase of

77.75 per cent. Though there was a slight decrease of 3.9 per cent and 11.2 per cent during the years 2002 and 2003, it recovered quickly, and increased to US$ 4.54 million in 2004 with 84.1 per cent increment from the previous year. Between 1997 and 2008, the government funding was an important source (mean 42.9 per cent, n=12) contributing to one third to over half of the overall budget (range 34.2 per cent -55.4 per cent) with the exception of 1998 (23.3 per cent, Figure 4).

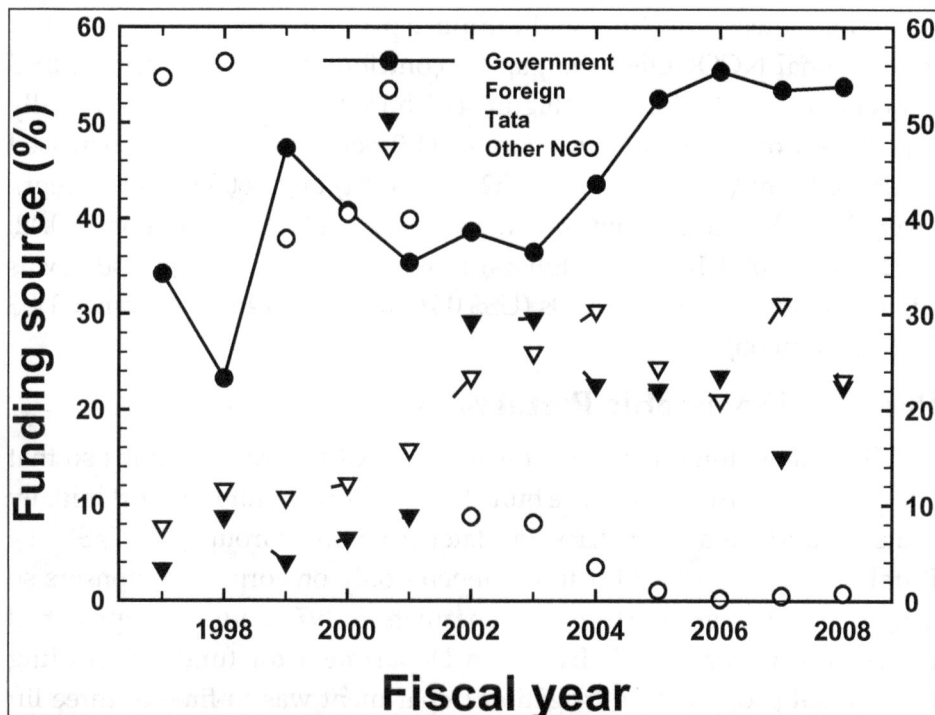

Figure 4: Relative Proportions (Per cent) of Funding Support Sadguru Foundation Received from Government, Foreign Sponsors, Tata Trusts and NGOs between 1997 and 2008.

The annual average government funding to the foundation was US$ 1.46 million (± 0.82, n=12) for a period of over 12 years. The second largest funding came from NGOs with an average of US$ 0.70 (± 0.46, n=12), followed by the corporate sector through Sir Ratan Tata Trust (SRTT) and Sir Dorabji Tata Trust (SDTT) (0.58 million ± 0.42) and foreign funding (0.45 ± 0.45). The government funding was significantly higher than other three sources (paired t-test, Government to Foreign: t=2.89, p<0.05; Government to Tata: t=5.75, p<0.01; Government to other NGO: t=5.75, p<0.01).

There was no significant difference within three sources (paired t-test, $P>0.1$). Between 1997 and 2001, foreign funding was the major source contributing nearly over half or one third (37.8 per cent -56.4 per cent) of the total (Figure 4) reaching up to US$ 1 million (0.8-1.15 million US$). But foreign funding dropped drastically to an average of US$ 101.7 ± 96.9 between 2002 and 2008 contributing to only 8.8 per cent (0.7 per cent) of the total budget (Figure 4).

Nonetheless, the financial support provided by Tata trusts and international NGOs filled the gap by contributing 44.5 per cent to 55.5 per cent of the total funding (Figure 4). Those two major sources initially contributed only US$ 179.5 (Figure 4, 11.0 per cent, 1997) but increased to US$ 2.40 million during 2004 (52.9 per cent) and 2005 (46.5 per cent). Overall, the Tata trusts contributed on average of US$ 0.58 million (± 0.42, n=12, range 0.05-1.14 million dollars) annually over a 12-year period. It was slightly lower than other NGOs (US$ 0.70 million ± 0.46, n=12, range US$ 0.13-1.37 million).

Pursuing Practicable Partnership

The major stumbling block for India's NGOs is to find funds so that adequate infrastructures can be built. Sadguru's corporate support initially came from Mafatlal Industries and later from Tata Group (SDTT/SRTT). Rural development NGOs cannot depend only on corporate sponsors so Sadguru had to seek help from other partners. In 1976, Sadguru approached the Gujarat government's Irrigation Department for funds. According to an initial proposal, the Irrigation Department was to finance three lift irrigation systems while Sadguru had to implement it. After completion, the community had to manage the structures. The government engineers wanted 6 months to prepare the plan, but Sadguru's engineers produced it within 2 weeks. Surprised at the speed, government officials reviewed the proposal and asked 60 technical queries, and all were adequately answered. Nevertheless, the Superintending Engineer objected the project based on the assumption that NGOs could not build irrigation structures due to lack of technical skills. The stand-off then reached the Special Secretary and the senior official stated that the government had wasted huge amounts for lift irrigation through numerous contractors therefore it would be worth supporting the small project (US$ 22,000). Finally, the Superintending Engineer approved the project and it was the first time that a government agency in Gujarat sanctioned a technical project to NGO. After the completion of three lift irrigation structures, the officials recognized the work as superior.

The initial government support to Sadguru started with the Special Secretary in Gujarat. Similarly, a senior official from Madhya Pradesh invited Sadguru in 1989 to start water resource work in Jhabua district. In the water-scarce Rajasthan, a high-ranking official sanctioned US$ 1 million in support of 45 projects in 1995 to build check dams, lift irrigation systems and percolation tanks in Banswara, Jhalawar and Chittorgarh districts. In this manner, Sadguru has not only established working partnership with the government but also influenced government policy, procedures and practices of rural development. When government enforced inefficient policies, the foundation criticized but it also appreciated the government while implementing worthwhile work (Jagawat, 2005). When Sadguru invited senior government officials, the government reciprocated by inviting the foundation to serve on its committees (Jagawat, 2005). Thus it has been involved in 13 district level committees in Gujarat, Rajasthan and Madhya Pradesh. It also became a member of the national level committee, 9 state-level steering and advisory committees. This shows that if NGOs perform neutrality with efficiency, they could become better partners. Sadguru's corporate partnership involved two firms; Mafatlal industries and Tata Group. The former established the foundation, and also continued funding (US$ 10,000-25,000) from 1974 to 1997. Another uniqueness of the Mafatlal and Tata is that they have supported poverty reduction projects without prejudicial interests. Many recent corporate sponsors in India tend to support projects mostly out of personal interest in their ancestral towns or near their factory units (Banerjee, 2007).

The Tata Group is India's largest with 114 companies and its founder, Jamsetji Tata had the vision to enhance India technologically proficient. His children, Dorabji Tata and Ratan Tata inherited the spirits. It's a tribute to that essence that 65 per cent of the capital of the parent firm, Tata Sons, is held by the trusts, SDTT and SRTT to serve humanity, which remains unique in India's corporate history (Sharma and Talwar, 2005). The SRTT was established in 1919 and Dorabji Tata left most of his personal wealth to SDTT before his death in 1932. In 1997, a senior manager from SDTT visited Sadguru and impressed with its poverty reduction work. As a result SDTT began funding, which continues till the time of writing. Similarly,

SRTT started its sponsorship since 2001. The understanding nature of senior managers of SDTT and SRTT towards Sadguru makes a resourceful partnership. Similarly, NGOs such as the Ford Foundation, Aga Khan Foundation and Norwegian Agency for Development (NORAD) supported Sadguru. Ford was the first foreign donor in 1988 followed by Aga Khan. NORAD's partnership started in 1991 with US$ 65,000 and increased to US$ 225,000 in subsequent years. The project ended after 13 years when India changed its policy towards foreign aid by allowing only USA, UK, Germany and Japan.

What is Unique in Partnership?

By maintaining credibility, high moral standards and absence of financial mismanagement, the Sadguru foundation was able to forge better working relations and trust of partners. Sadguru is an accredited member of the Credibility Alliance. The foundation's advisory board consists of several senior officials from government agencies, corporations and NGOs that provide funding. This forum provides an opportunity to donors to review the development work to make it transparent. Although Sadguru does not include community members in its advisory board, it provides village institutions as platform for donors to interact with community leaders and members. This way, the donors had better access to local community directly through village institutions to assess the ground reality and impact of development work. Although Sadguru played a crucial role in initiating and maintaining partnership with government, corporations and NGOs, the communities in fact formed the backbone in the partnership endeavor, and without it, development work could not have succeeded at the grassroots. Representatives from the government, corporate and NGO sectors interacted with community members of village institutions on a quarterly/bi-annual basis and Sadguru liaised to strengthen communications between sponsors and community leaders/members. Recording the minutes of such meetings that brought together the four sectors on one platform to discuss strategies, enhance rural development, and reduce poverty in villages were kept by community leaders as well as Sadguru foundation. Based on those records, new development initiatives and proposals for funding were initiated. For example, after the completion of check dams in villages, lift irrigation systems were built. Without water from check dams, lift irrigation systems could not have been built. The hydro-centric schemes subsequently supported horticulture, floriculture, vegetable farming, social forestry, and join-forest management leading to livelihood improvement and poverty reduction over 1.5 million people across India's drylands.

Due to the expertise on water resource and mobilizing local communities, Sadguru became a resource center for rural development planning initiatives of the government of Gujarat, Rajasthan and Madhya Pradesh. They continue to collaborate with Sadguru while preparing rural, tribal and water resource development plans for the region each year. Sadguru provides trainings to government agencies in the fields of irrigation, water management, and agriculture. For example, leading civil administrators from the Indian Administrative Service (IAS) took part annually in trainings at the foundation to understand poverty reduction schemes. Besides, officials of all categories including the National Bank of Agriculture and Rural Development (major sponsor of development work) participated in trainings at the foundation's headquarters each year. Sadguru has a fully-equipped training center with guest houses, dormitories, conference halls and meeting rooms that provide better opportunity for hands-on learning about development work. Numerous training programs have been conducted over the years for various government agencies comprising nearly 21 states across India.

Conclusion

When the government, corporate and NGO sectors funded water resource projects, they relied on commitment, proven track record in community mobilization and compliance of local/national regulationsSadguru had those credentials enhancing partnership. But the fundamental for partnership is to get the support of communities and Sadguru succeeded by empowering people via village institutions and self-help groups. The strong community support boosted credibility and also tackled political problems. For example, on two occasions, politicians forced government officials not to sanction rural development projects to Sadguru foundation. But, it provoked public protests that instantly forced politicians to back away. Although funding from donors was used to build water resource structures, Sadguru had made it clear to its partners that communities would have the ownership. As a result, communities continue to use and maintain the well-built water harvesting structures through village institutions. Due to strong community involvement, water resource development work became successful. On the contrary, thousands of check dams and lift irrigation systems constructed by the Indian government through private contractors have failed due to lack of community involvement and flawed construction (Choudhry *et al.*, 2002). Similarly, the government of China had built 100,000 check dams during the 1960s and 1970s to reduce sediment flow and to improve irrigation, but 80 per cent of dams built in Shanbei region had failed. Lack of community participation and poor construction caused the failure (Xu *et al.*, 2004).

The nature of water resource development work done at the grassroots by a non-profit agency and the nature of learning obtained from the community to reduce poverty provide opportunities for social workers not only to learn more on social work but also to use the learned knowledge to expand social work practice in the poverty reduction context. This study has shown that even a small NGO can work with community in partnership with government, corporate and NGO sectors and it also provide opportunities for social workers not only to enhance social work approaches but to use applied knowledge to implement rural development and poverty reduction targets. This approach has the potential to be tested by other NGOs across the developing world.

References

Ackerman, S.R. (2008). Corruption and government. *International Peacekeeping*, 15, 328-343.

Agoramoorthy, G. (2008). Can India meet the increasing food demand by 2020? *Futures*, 40, 503-506.

Agoramoorthy, G. (2009). *Sustainable development: The power of water to easy poverty and enhance ecology*. Delhi: Daya Publishing House.

Anand, R. and U. Anand, U. (2007). India needs its NGO. *Harvard International Review*, 8, January.

Ashman, D. (2001). Strengthening North–South partnerships for sustainable development. *Nonprofit Volunteer Sector Quarterly*, 30: 74–98.

Austin, J. E. (2000). Strategic collaboration between nonprofits and businesses. *Nonprofit Volunteer Sector Quarterly*, 29: 69–97.

AVARD (1991). *Role of NGOs in development: A study of the situation in India: Final Country Report*. Delhi: Association of Voluntary Agencies for Rural Development.

Backstrand, K. (2006). Democratizing global environmental governance? Stakeholder democracy after the World Summit on Sustainable Development. *European Journal of International Relations,* 12, 467-498.

Banerjee, S. B. (2007). *Corporate social responsibility: The good, the bad and the ugly.* Bostoms: Edward Elgar Publishing Inc..

Boris, E. T. and C. Eugene, S. (1999). *Nonprofits and government- collaboration and conflict.* Washington DC: UIP.

Brinkerhoff, J. M. (2002). Government–nonprofit partnership: a defining framework. *Public Administration and Development,* 22, 19-30.

Central Statistical Organization (2010). *NGO Statistics.* Delhi: Ministry of Statistics and Programme Implementation.

Choudhry, K. *et al.* (2002). A Study of government-installed lift irrigation schemes in district Jhabua, Madhya Pradesh. Dahod: Sadguru Foundation Report.

Cohen, J. M. and N. T. Uphoff (1977). *Rural Development Participation: Concepts and Measures for Project Design, Implementation and Evaluation,* Ithaca: Cornell University.

Dewan, P. K., S. S. Lal K. Lonnroth, F. Wares, M. Uplekar, S. Sahu, R. Granich and L.S. Chauhan (2006). Public-Private Mix in India: Improving Tuberculosis Control Through Intersectoral Partnerships. *British Medical Journal,* 332, 574-578.

Government of India (1992a). *Eighth Five Year Plan, 1992-97. Volume I, Objectives, Perspectives, Macro-Dimensions, Policy Framework and Resources.* Delhi: Planning Commission.

Government of India (1992b). *Eighth Five Year Plan, 1992-97. Volume II, Sectorial Programmes of Development.* Delhi: Planning Commission.

Government of India (2001). *Census of India 2001.* Delhi: Government of India Press.

Heap, S. (2000). NGO-business partnerships: Research in progress. *Public Management,* 4, 556-63.

Jagawat, H. (2005). *Transforming the dry lands: The Sadguru story of western India.* Delhi: India Research Press.

Jain, P. (1989). *Perspectives on Voluntary Organization.* Anand: IRMA.

Kothari, R. (1986). NGOs, the state and world capitalism. *Economic and Political Weekly. 21*, 2177-2182.

Lister, S. (2000). Power in partnership? An analysis of an NGO's relationships with its partners. *Journal of International Development, 12*, 227-239.

Mehra, A. K. (2000). Naxalism in India: Revolution or terror? *Terrorism and Political Violence, 12*, 37-67.

Midgleya, J. and M. Livermore (1998). Social capital and local economic development: Implications for community social work practice. *Journal of Community Practice, 5*, 29-40.

Mikkelsen, B. (1995). Methods for development work and research: A guide for practitioners.New Delhi: Sage.

NCEUS (2007). *Report on the conditions of work and promotion of livelihoods in the unorganized sector*. Delhi: National Commission for Enterprises in the Unorganized Sector.

NCRB (2007). *Accidental death and suicide in India*. Delhi: National Crime Records Bureau.

OECD (2001). *DAC Guidelines on poverty reduction*. Paris: Organization for Economic Co-operation and Development.

Price Waterhouse Cooper (2002). *Putting partnering to work*. Washington DC: World Bank.

Rajasekhar, D. and R.R. Biradar (2004). *Reluctant partners coming together? Interface between people, government, and the NGOs*. Delhi: Concept Publishing Co.

Rangachari, R., N. Sengupta, R. R. Iyer, P. Banerji and S. Singh (2000). *Large dams: India's experience*. Cape Town: World Commission on Dams.

SAS Institute (2000). *SAS/ETS software: changes and enhancements*. Cary: SAS Institute.

Sen, S. (1999). Some aspects of State-NGO relationships in India in the post-independence era. *Development and Change, 30*, 327-355.

Sharma, A. K. and Talwar, B. (2005). Corporate social responsibility: Modern vis-a'-vis Vedic approach. *Measuring Business Excellence, 9*, 35-45.

Varshney, A. (2000). *Ethnic conflict and civic life: Hindus and Muslims in India*. New Haven: Yale University Press.

Warner, M. (2002) *Optimizing the development performance of corporate investment.* London: Overseas Development Institute.

Warner, D. and R. Sullivan (2004). *Putting partnerships to work: Strategic alliances for development between government, the private sector and civil society.* Sheffield: Greenleaf.

World Bank (1997). *World development report 1997: The state in a changing world.* Washington, DC: World Bank.

World Bank (2001). *World Development Report 2001: Attacking poverty.* Washington, DC: World Bank.

Xu, X. Z. *et al.* (2004). Development of check-dam systems in gullies on the Loess Plateau, China. *Environmental Science and Policy, 7,* 79–86.

Young, D. R. (2000). Alternative models of government–nonprofit sector relations: theoretical and international perspectives. *Nonprofit and Voluntary Sector Quarterly, 29,* 149-172.

Zaidi, S. K. (1999). NGO failure and the need to bring back the state. *Journal of International Development,* 11, 259-271.

Chapter 2

Community Farming Promotes Food Security

"Eating is an agricultural act."

— *Wendell Berry*

Introduction

The Food and Agriculture Organization states that farmers are the largest investors in the developing world agriculture sector so they must actively participate to play a central role while governments develop future strategies (FAO, 2012). The arable land worldwide has been estimated to be high reaching up to 3.339 billion acres, but it only coms to 0.49 acres per person (Anderson and Lafond 2010). Asia has the largest stake of the arable land in the world (32 per cent), which is followed by North America (17 per cent) and Africa (14 per cent). Besides, India leads the world in harboring the largest irrigated area with 96.4 million acres, which is followed by China with 46.9 million acres, and USA with 42 million acres, respectively (Renner, 2012). Agriculture productivity largely depends on irrigation water. So the irrigation sector consumes nearly 70 per cent of all the freshwater withdrawals worldwide destabilizing the natural ground/surface water supply and demand cycle (Jarvis, 2014).

Irrigation has been known to offer better yields of various types of crops and grains across India. They are about four times more than the traditional rain-fed farming. However, the irrigation expansion has shown a slowdown in India since the 1970s due to various problems including the decline in irrigation investment to poor performance of large canal irrigation systems combined with corruption and mismanagement in the infrastructural construction processes (Hussain and Hanjra, 2004). India has also experienced rapid economic development often at the cost of natural environment (Bajpai, 2000). Ground water, which is crucial for agricultural development, has been severally depleted. Scientists argue that India's green revolution has gone brown due to the creation of agrarian class differentiation, poor soil quality, ecological degradation, decreasing yields and falling groundwater (Greenland 1997; Lipton and Longhurst 1989; Shiva 1992; Atkins 2001; Jha 2002; Kapila and Kapila 2002; Jagawat 2005; Sharma 2007; Agoramoorthy 2009).

The downsides of India's irrigation and agriculture strategy are the historic neglect of catchment areas in remote drylands where tribal communities (also known as *adivasi*, meaning 'original people') inhabit for centuries. India's tribal region incidentally is the poorest in the country since the tribal communities have scattered across the remote wilderness where access to public services is minimal (Phansalkar and Verma, 2005). From the government point of view, a single tribal settlement is too small to economically justify a school or a health center, and the poor infrastructure makes other facilities elsewhere difficult to access. Furthermore, the low density of the tribals and their lack of purchasing power make the village level business not at all feasible (Janaiah *et al.*, 2000).

Within the rankings of the Indian traditional caste system, tribals are beneath the untouchables, thus the most downtrodden socially and economically. The Indian constitution of 1950 singled them out for preferential treatment, in a kind of permanent affirmative action plan, but the government efforts to reduce poverty in tribal areas did not fully succeed for over half century (Jagawat, 2005). The economy of tribal communities often revolves around the interactions between forest, agriculture and migration (Agoramoorthy, 2009). Therefore, if the government plants to promote sustainable agriculture strategies in rural India, it simply cannot neglect the tribal communities.

Unlike the non-tribal communities that prefer to live in the plains and coastal areas, India's tribal societies often dwell in the semi-arid uplands and riverbanks (Phansalkar and Verma, 2005). The drylands receive 400 to 1000 mm of rainfall. Due to undulating terrain influenced by the lack of adequate water harvesting structures, much of the rainwater runs off and is not fully utilized for agriculture. The commonly used flow irrigation or gravity irrigation in flatlands is not suitable for drylands due to rugged terrain where farms are at a higher level than the water source (25-40 m, Table 1). Therefore, lifting water from its source to upland farms using mechanized pumps known as 'lift irrigation' is the best irrigation option for drylands (Jagawat, 2005). The lift irrigation needs to be supported by water harvesting structures such as lakes, rivers, percolation tanks and check dams. The check dam is a small barrier built across the direction of water flow on small, medium and large rivers to harvest rainwater to be used for irrigation. In order to build the lift irrigation and check dam structures, funding is crucial hence partnership involving government, non-government and private corporations is necessary. Then only, rainwater could be harvested, stored and distributed to farms. In fact, past studies have shown that how irrigation could reduce rural poverty (Jagawat 2005; Chitale 1994; Brisco 1999; Bhattarai *et al.*, 2007).

Comprehending the irrigation problem of drylands, the Sadguru Foundation has pioneered a community-oriented approach in harvesting, storing and distributing rainwater with minimum wastage to improve agriculture (Jagawat 2005; Agoramoorthy and Hsu, 2013). Although participatory approaches have developed in popularity over the last few decades to enhance positive change in marginalized communities, they are now increasingly used to deal with the complex issue of sustainable development (Bruges, 2008). Despite globalization, the governments' efforts to assist farmers in developing countries through agriculture extension and subsidies have not succeeded fully to meet the harsh realities of farming (Shiva, 1992). Therefore, the governments of developing countries have turned to participatory approaches to make agriculture more ecologically and economically sustainable (Pretty 1995; Singh and Ballabh 1996; Jagawat 2005). However, understanding the potentiality of sustainable agriculture requires thorough analysis of the nature, purpose, problem and prospect of community participatory approaches (Bruges 2008).

In this chapter, I have analyzed a case study on community lift irrigation scheme implemented in India's drylands to answer the following questions.

Table 1: Average (± SD) Height, Pumping Capacity (Horse power), Irrigated Area (Acre) and Beneficiaries of Lift Irrigation Schemes (n=382) implemented by Sadguru Foundation in 10 Districts of Gujarat, Rajasthan and Madhya Pradesh States of India

District Name	n	High (m) Mean±SD	Area (Acre) Mean±SD	Pumping capacity Mean±SD	No. of families Mean±SD
Banswara	63	40.7±13.6	166.9±39.5	37.8±18.5	51.0±22.8
Dahod	214	37.4±13.8	300.0±152.0	51.3±32.3	76.3±46.1
Dhar	1	67.0±0.0	256.0±0.0	45.0±0.0	58.0±0.0
Dungarpur	15	41.9±6.7	123.4±33.3	26.7±6.7	28.3±12.4
Jhabua	16	32.9±8.3	275.9±108.6	43.1±22.4	85.1±30.3
Jhalawar	40	43.6±15.7	235.9±53.2	50.5±24.3	68.8±25.8
Mandsor	1	67.0±0.0	450.0±0.0	120.0±0.0	135.0±0.0
Panchmahal	24	37.4±17.5	326.7±144.9	47.0±18.0	71.4±40.3
Ratlam	2	77.0±19.8	268.0±130.1	75.0±21.2	73.0±21.2
Vadodara	6	22.7±11.4	307.3±78.8	22.5±12.5	24.8±4.1
Mean	382	38.7±14.5	264.9±137.2	47.3±28.5	69.0±41.4

Is there any way to mobilize local communities to harvest, store, distribute and use rainwater efficiently for irrigation? What are the advantages of using community participatory approach to manage lift irrigation schemes? Can the community lift irrigation approach increase irrigated area, expand the number of beneficiaries, enhance crop yield and relive poverty? What are the potentials of community irrigation to mitigate future food shortages and farmer suicides? I have done a systematic scientific analysis on participatory lift irrigation schemes to answer the above questions.

Study Area

The study site description for districts namely Dahod in Gujarat State, Jhalawar and Banswara in Rajasthan state and Jhabua in Madhya Pradesh state where most lift irrigation systems were established between 1976 and 2012 by a non-profit organization Sadguru Foundation are outlined below. The Dahod district (area 3,642 km^2) harbors a population of about 1.6 million and is one of the poorest in Gujarat. Most people (72 per cent) belong to the Bhils tribe. The region receives 860 mm of annual average rainfall resulting in drought on average every third year. The Banswara district (area 5,037 km^2) is inhabited by the Bhils, Bhil Meenas, Damor, Charpotas, and Ninamas tribes (Government of India, 2011). The district has a population of about 1.5 million of which 94 per cent in rural areas. Agriculture is the main occupation, and tribal families live in small houses (locally known as *tapra*) indicating poverty. The Jhalawar district (area 6928 km^2) is not only the poorest in Rajasthan but also one of the least developed in India. It supports a population of about 1.2 million and vast majorities (86 per cent) live in rural areas (24). The average annual rainfall is about 950 to 1000 mm. The Jhabua district in Madhya Pradesh State (area 6,782 km^2) supports a population of 1,396,677. About 85 per cent of people belong to tribal communities. The overall literacy rate is 36.87 per cent with female literacy of only 4 per cent (Government of India, 2011).

Data Collection and Analysis

Each lift irrigation unit includes components such as pump house to keep machinery, distribution system with underground pipelines, main delivery chamber located at the highest point where water is lifted, and electrical accessories. The transformer and power supply are provided by the government-run electricity board. The underground pipelines form a network in which water flows by gravity and are connected to distribution chambers. Villages located in Gujarat, Rajasthan and Madhya Pradesh states were visited between August 2008 and December 2012 to collect data on lift irrigation systems in check dams, reservoirs, rivers, canals, and tanks, and their irrigation benefits following methods described by Creswell (1994). A total of 391 lift irrigation systems were completed between April 1976 and March 2014 of which 382 (Table 1) are included in statistical analysis. Data on the year of operation, water source, motor pumping capacity, lifting heights, total cost, expansion of irrigated areas and number of beneficiaries were pooled from the archives of Sadguru Foundation. Discussions and interviews of 330 people were conducted while visiting lift irrigation corporatives in villages to record the impact in terms of increased irrigation area, improved agricultural productivity, and enhanced livelihood, following the methods of Mikkelsen (1995).

Major monsoon crops (*Kharif* in Hindi) are maize and rice that are grown from mid-June to late September while the post-monsoon crops (*Rabi* in Hindi) that include wheat, maize, gram, and pulses are grown from mid-October to mid-February. Data on the improvement in housing (traditional, semi-standard and standard) before and after lift irrigation were collected from six villages. Traditional houses are built with mud, wood and rock (without concrete) and covered by thatched roof. The semi-standard houses are built by a combination of brick, cement, rock, mud and wood, which is stronger than traditional houses. The standard houses are built by bricks and cement with concrete foundation; they are the strongest. Farmers' suicide data were pooled from the archives of the National Crime Records Bureau (NCRB, 2014). All statistical analyses were conducted using the Statistical Analysis System software (SAS Institute, 2000). All mean values are presented as ± 1 standard deviation. Local currency (India rupee) was converted to US$ according to yearly exchange rate of the Indian government. Pearson Correlation was used to test the relationships among lift height, capacity, area of lift irrigation, family benefited and the construction cost. Varies general linear models were used with analysis of variance to test the effects of variables such as year, lift height, capacity, area of lift irrigation, and family benefited on the construction cost. A general linear regression was used to estimate costs for the above variables (SAS Institute, 2000).

Community Irrigation

The community lift irrigation is managed by the user farmers in villages (Figure 5). The cost to construct lift irrigation units and check dams was sponsored by government agencies responsible for rural development through Sadguru Foundation with matching funds from private corporations. Tribal people who inhabit India's drylands own lands with an average holding of 2 acres, which gives them farming opportunity (Jagawat, 2005). The objectives of creating lift irrigation cooperatives in villages were to increase capacity building of farmers to enhance self-sufficiency in food production. Emphasis was given on community-oriented collective management of water and natural resources to promote sustainable livelihood and financial self-reliance. Peoples' involvement was ensured from the beginning of a scheme. It began with visits to villages, meetings of

Figure 5: Lift Irrigation Along a River Bank in Gujarat, India (Above) and Water Lifted Up Diverted to Field through Gravity (Below) (Photo courtesy Sadguru Foundation).

farmers, and discussions to raise awareness on issues surrounding collective management of irrigation opportunities.

Once a farmer became a member of the community lift irrigation cooperative, he/she was eligible to get irrigation water from the participatory system. Irrigation cooperatives are generally managed by a committee of 12 elected members. The committee is headed by a chairman and assisted by a secretary, who keeps financial records and organize monthly meetings to inform members of activities.

Apart from holding meetings, the committee oversees auditing, financial management, water distribution, collection of water charges, payment of electricity bills and staff salaries, maintaining or repairing lift irrigation systems, and solving water distribution disputes. When some farmers are unable to pay their dues, they are allowed to pay in full before the next irrigation season with an annual interest of 20 per cent. Then only the farmer can be entitled to receive water. All cooperatives save money and profits are deposited in bank, which shows the self-sufficiency in irrigation. It also shows how farmers managed village-level institutions with efficiency and self-reliance.

Expansion of Irrigated Areas

Between 1976 and 2012, 382 lift irrigation systems were constructed in 10 districts such as Dahod, Banswara, Jhalawar, Panchmahal, Jhabua, Dungarpur, Vadodara, Ratlam, Mandsor and Dhar covering Gujarat, Rajasthan and Madhya Pradesh states (Table 1). Dahod district had 56.0 per cent of the systems; it was followed by Banswara (16.5 per cent) and Jhalawar districts (10.5 per cent; Table 1).

These districts are classified as drought prone drylands and inhabited predominantly by the impoverished tribal communities. The average height of lift irrigation system was 38.7 ± 14.5 m (n=382) and the average amount of benefited irrigation area was 264.9 ± 137.2 acres (Table 1). The average pumping capacity of lift irrigation was 47.3 ± 28.5 horse powers and the average number of beneficiaries was 69.0 ± 41.4 households (Table 1).

About US$ 18.59 million was spent to construct 382 lift irrigation systems that enabled a total water pumping capacity of 18,002.5 horsepower from 1976 till 2012. The lift irrigation system in general expanded the irrigated area to 100,415 acres (Figure 6). On average, 11.6 ± 7.1 (n=33) lift systems were constructed from 1976 to 2012. Between 1990 and 2012, an average of 14.70 ± 5.97 lift systems were constructed yearly with maximum constructed in 2004 (n=31) and minimum in 1993 and 2012 (n=5), respectively. The average cost to build the lift system per year was US$ 563,330 ± 3.55 (n=33).

The height of lift, pumping capacity, irrigated area and year had effects on the cost of lift construction (F36, 334=39.56, R2=0.81, p<0.001). Among them, irrigated area was the major factor (F1, 334=22.0, p<0.001). The construction costs were positively correlated to lift height, pumping capacity, area of lift irrigation and number of family benefited (Pearson Correlation, p<0.001, Table 2). The average of the accumulated height of lift irrigation constructed each year was 442.5 m ± 298.7 (n=33), with the highest in 2004 (1311 m).

Table 2: Pearson Correlation Coefficients of 382 Lift Irrigation Systems Implemented by Sadguru Foundation in Gujarat, Rajasthan and Madhya Pradesh States of India

	Lift Height	Cost	Area	No. of Family
Pumping capacity	0.53***	0.70***	0.66***	0.62***
Lift height	1	0.46***	-0.02	0.05
Cost		1	0.51***	0.52***
Area			1	0.76***
No. of family				1

* p<0.05, ** p<0.01, *** p<0.001.

It was followed by 2007 (927 m), and 2011 (857.5 m), respectively. The accumulated cost of constructing lift irrigation was highest in 2011 (US$ 1.38 million; Figure 7), which is about 3.07-3.24 times that of 2008 (US$ 0.42 million) and 2012 (US$ 0.45 million). However, the lift cost per meter was lowest in 2006 (US$ 735.8/m), highest in 1983 (US$ 2554.8/m; Figure 8). In addition, the cost of benefited irrigation area per acre was lowest in 1975 (US$ 63.33/acre) and highest in 2009 (US$ 498.2/acre; Figure 8). The lift irrigation did not have any negative impact to natural environment.

The area irrigated after the construction of lift irrigation was highest in 2004, followed by 2007 (Figure 6). The highest amount of irrigated area increase was 6625.5 acres in 2004 from 31 lift construction completed with a total pumping capacity of 1355 horsepower, followed by 5512.5 acres (2007) from 23 lifts completed with a total capacity of 1110 horsepower (Figure 6).

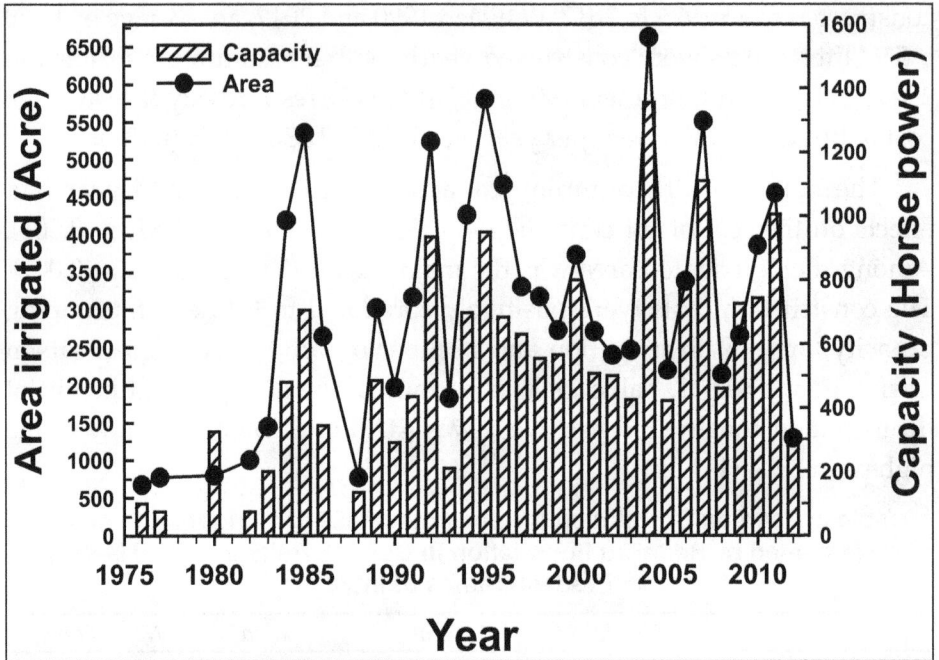

Figure 6: Total Irrigated Area and Water Pumping Capacity of Lift Irrigation Systems Constructed between 1976 and 2012 in India.

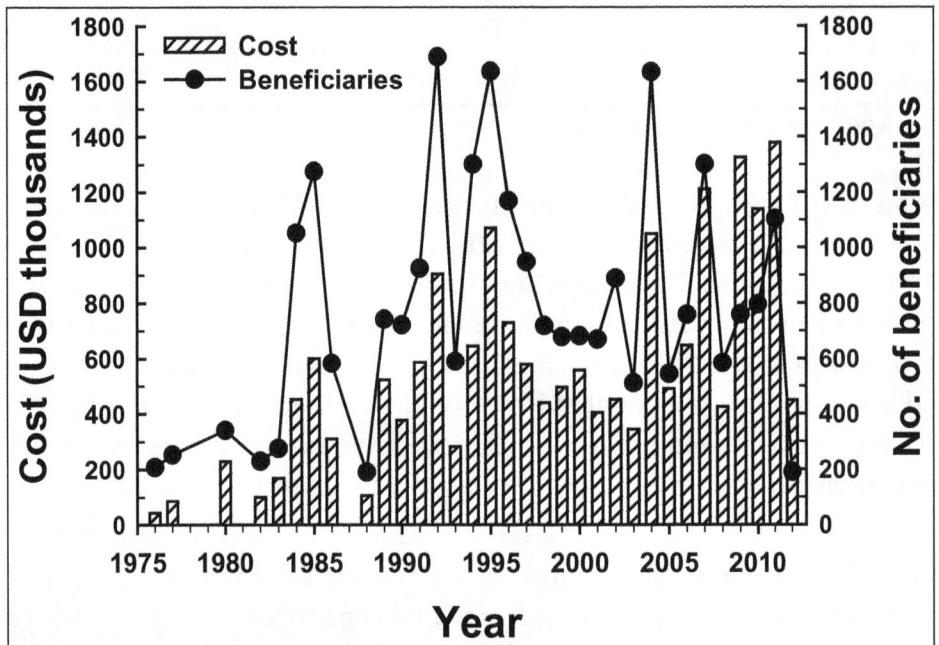

Figure 7: The Cost (US$) and Number of Additional Houses Benefited each Year of Lift Irrigation Systems Constructed between 1976 and 2012 in India.

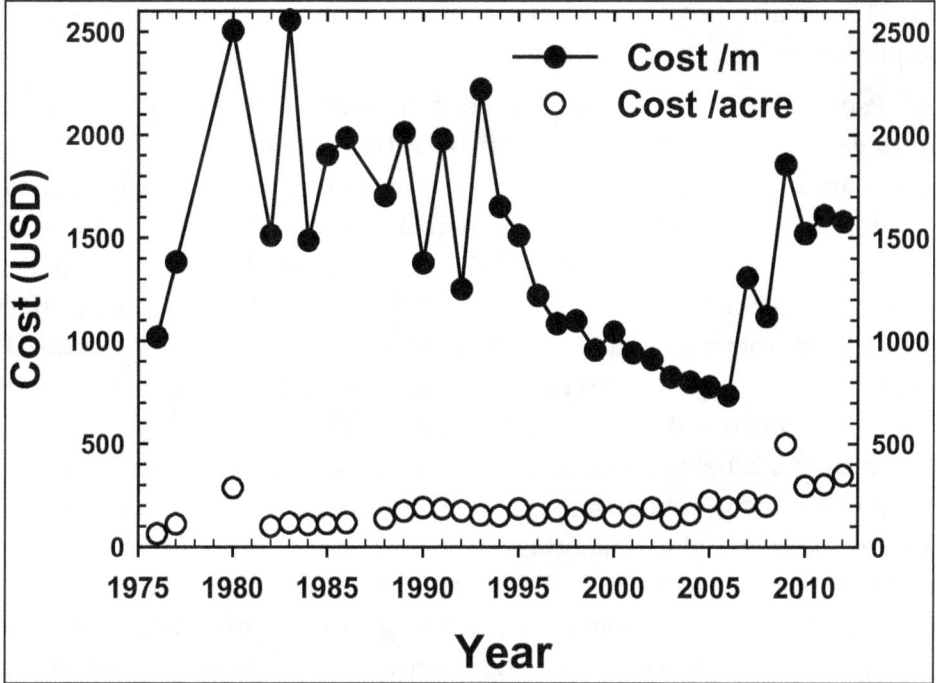

Figure 8: The Cost (US$) of Lift Irrigation per meter Height or Area Irrigated per Acre Constructed between 1976 and 2012 in India.

The number of beneficiaries was highest in 1992 with 1688 families (Figure 6), followed by 1995 (1636 families) and 2004 (1634 families). Each household supports an average of six members. They traditionally grew one rain-fed crop yearly and frequent droughts often forced them out of their villages in search for jobs in nearby towns and cities (Jagawat 2005; Agoramoorthy 2009). However, after the implementation of lift irrigation schemes, farmers migrating to nearby towns and cities in search of jobs stopped and agricultural productivity increased leading to the elimination of poverty.

Crop Productivity and Livelihood Upgrading

Data on crop productivity in six villages namely Parasali, Khejadia, Motizer, Zerjitgadh, Motadharola and Sampoi before and after lift irrigation were analyzed to show how it enhanced crop production. The change in the crop yielding pattern was highest in post-monsoon season for grams and pulses (85.1 per cent), from 96.3 kg/acre (± 88.2) increased to 648.17 kg/acre (± 370.3). Similarly, the post-monsoon maize yield after lift irrigation dramatically increased to 79.2 per cent, from 140 kg/acre (± 198) increased to 671.5 kg/acre (± 449.0). Even the monsoon crops (soybean and rice) that require more water also increased in yield.

People lived in traditional houses built with mud and thatched roof prior to the implementation of lift irrigation schemes in villages. But, afterwards the percentage of traditional houses in six villages such as Zerjitgadh, Parasali, Khejadia, Motadharola, Motizer and Sampoi decreased dramatically from 94.4 per cent (n=125) to 51.2 per cent. The average number of traditional houses in six villages was 19.7 (± 8.6) and it decreased to 10.7 (±4.1). On the other hand, the percentage of semi-standard houses increased nearly nine times (4 per cent to 35.2 per cent) while the strongest standard houses also increased 8.5 times (1.6 per cent to 13.6 per cent). This indicates the economic benefits obtained from lift irrigation schemes contributing to the improvement in housing standards in villages. All households in the above villages faced food shortages each year due to droughts and lack of irrigation water. So people relied on migration to nearby towns in search of jobs. After the implementation of lift irrigation schemes, all households attained self-sufficiency in food grain production that lasted for the whole year. Hence the seasonal migration of farmers in search of jobs in nearby towns had stopped. It was due to increase in irrigated area to produce crops that were not possible before.

Community Irrigation and Dairy Farming

India is one among the largest milk producers in the world. It produced 132.4 million t during 2012-13, which is 15 per cent of the global production (Sood 2014). Of which, 87,824.21 t of dairy products were exported.

Nonetheless, little is known on how the lift irrigation plays a role in dairy farming. A village named Mota Darola in Gujarat has an outstanding example. The village supports a total of 102 households that also hold 250 buffaloes and 50 cows; each animal costing about US$ 825. Although the village had access to canal irrigation in the past, water did not reach all farms due to uneven terrain. The low groundwater further discouraged even those who thought about drilling. Understanding the terrain-induced setbacks, farmers had approached the Sadguru Foundation in 1996 with a request to set up a lift irrigation scheme since it had already been proved as an ideal technology for the rugged terrain (Jagawat, 2005). After the implementation of the lift irrigation scheme, irrigation water reached all farms (area 470 acres) in the village.

The self-sufficiency in irrigation has led to better production of crops, vegetables and fruits and even fodder for livestock. The 90-member village dairy cooperative on average started to produce 1100 liters of milk daily (Figure 9). Therefore the community's total annual profit from milk alone came to US$ 165,000. Even if they get half of the profit after deducting overheads, it is an enormous benefit for a small village. The village started its first dairy business in 1987, but the number of members and milk production started to increase only after the lift irrigation scheme (Figure 9). The milk output has more than doubled over the last 14 years showing the long-term impact of lift irrigation that provides the needed water to irrigate farms ultimately leading to the production of more fodder for livestock.

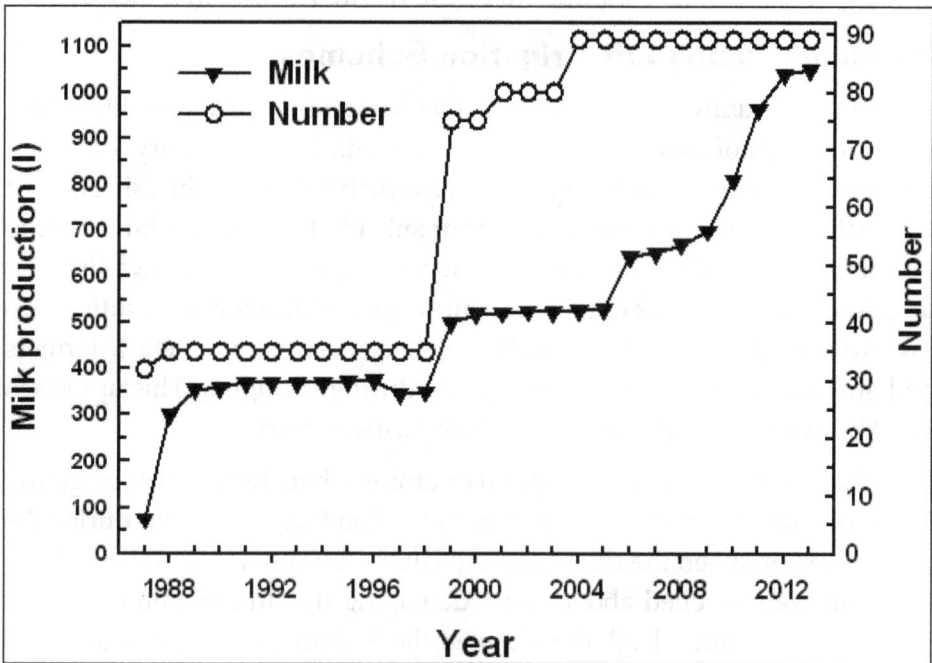

Figure 9: Increase in Membership of Dairy Cooperative and Average Daily Milk Production (litres) between 1987 and 2013 in Mota Darola village (Gujarat).

The community members treat their buffaloes and cows dearly so they stay with their human families throughout their lives, even after they become unproductive. What fascinated the authors was that people keep livestock in their living rooms where both humans and non-human companions watch television together. Moreover, some even built concrete houses to keep their livestock showing economic incentives brought about by them. The dairy cooperative even owns a high-tech freezer to store milk where trucks come regularly to pick up, transport and distribute the commodity. Besides, they have veterinarians who routinely watch over the health of livestock. These benefits extended by the village-level dairy cooperative cannot be accessed by individual farmers (Mascarenhas, 1998). The above mentioned village-based diary-cooperative model also reflects the national trend on how the so called "white revolution" that transformed India's past milk-deficient status prior to the 1970s, to one of the world's leading milk producers, all due to the influence of the dairy cooperative scheme that originated in Gujarat during the 1940s (Kurien 2004). So the country's dairy revolution may go on without glitches because India continues to be the world's largest livestock owner.

Problems Facing Lift Irrigation Schemes

The community lift irrigation schemes sometimes encounter difficulties involving theft of machinery, inability to obtain lost machinery, disputes, political interferences, inability to pay electricity bill, droughts, shortage of electricity, and lack of crop insurance or subsidy to farmers who face crop failure or natural disasters. The situation in Gujarat and Madhya Pradesh is serious due to the lack of government support to farmers who suffer crop losses during droughts. However, Rajasthan has been sympathetic to farmers and subsidizes or cancels electricity bill during droughts. This approach must be extended to other states to help farmers in crisis.

Flood is one of the major natural calamities that affect the lift irrigation infrastructure and machineries. For example, flooding was serious during the 2012 monsoon when the rivers namely Hiren (Rajasthan), Hadaf and Anas (Gujarat) were swelled above shore damaging the lift irrigation systems. Then the community leaders requested the Sadguru Foundation to assess damage. Experts were sent to collect data and to work out repair strategy. The analysis was done within a week that pointed damages to civil structures (pump house and well), mechanical structures (pipes-succession, delivery, valves), and electrical structures (pump, motor, starter, panel board). After a meeting, the leaders of lift irrigation federations decided to cover 25 per cent of the repair costs. Within a month, all 72 damaged lift irrigation systems were repaired and crops were irrigated without interruption. In the meantime, Sir Dorabji Tata Trust (SDTT) had allocated 80 per cent of funds while the rest of the repair expenditure was covered by the farmers' irrigation federations. This shows the efficiently of village-level cooperatives in tackling the crisis timely. But the government-managed lift irrigation systems need to wait for long time since it is up to the courtesy of officials to respond to disasters. If houses are damaged or livestock killed during natural disasters, the government can pay for the loss rather quickly. But repairing community irrigation structures is not all that imperative from the government point of view to deal the crisis quickly. That's why most government-built lift irrigation schemes seldom function (Choudhry *et al.*, 2002).

Farmer Suicide and Community Irrigation

About 135,445 cases of suicide in 2012 were recorded across India (NCRB 2014). The self-employed category accounted for 38.7 per cent of victims, of which 11.4 per cent were engaged in farming. Age-wise profile of victims shows that nearly 36.7 per cent of them were farmers (30-44 years of age). India has recorded over a quarter million farmers committing suicide between 1995 and 2010. In 2010 alone, 15,964 farmers committed suicide and it brings the cumulative 16-year total from 1995 when data collection started to 256,913, which is the worst-ever recorded farmer suicide in the history (NCRB 2014). Due to India's recent economic outbursts, the cost of formers' basic needs has gone up while earnings are sinking due to the increase in price of seeds, pesticides and fertilizers. When farmers are faced with credit squeeze by legitimate banks, they are forced to borrow money from illegal loan sharks. After they are engulfed by debt due to repeated crop failures, suicide becomes the only salvage. But the government is aware of the fact that debt-ridden poverty is driving farmers towards suicide. Nevertheless, it has not come up with a workable strategy to provide sufficient emergency support funds to cover the cost of crop production ahead of the worst case scenarios involving crop failures (Gaiha 2000; Agoramoorthy 2008).

Although the crisis seems to be unmanageable, it can be tackled if farmers form community-based approach to deal irrigation shortages as shown in this paper. When farmers face financial hardships, the final safety net constitutes their friends, family and community. The formation of irrigation cooperatives can serve as 'social capital', which can be used by the families, friends, and associates of the impoverished farmers at times of crisis. When communities form social networks similar to the irrigation cooperatives discussed in this paper, they are in a comfortable position to confront poverty, escape from suicide (North 1990; Narayan 1995), resolve social disputes (Varshney, 2000), and ultimately take advantage of rural development economic opportunities (Isham, 2000). Therefore, poverty and suicide can be naturally neutralized by reviving participatory approaches at the grassroots in villages.

As a matter of fact, the participatory approach is not new for India, and it was first introduced in 1904 when the Cooperative Credit Societies Act was ratified. Later the act was amended in 1912 to include non-credit institutions and federal organization (Singh and Ballabh, 1996). Besides, the government enacted the Multi-state Cooperative Societies Act in 2002 to provide democratic and autonomous working of cooperatives. The village-level irrigation cooperatives highlighted in this paper are united by state-level federations and registered under the cooperative societies act. When people collaborate to create their own social rules, opportunities for individuals and collective empowerment can emerge (Ostrom 1992; Singh *et al.*, 2001). Participatory approaches to manage natural resources including irrigation, forestry, salt mining, watershed, and fisheries were known to not only improve livelihoods, but also minimize suicides (Singh and Ballabh , 1996; Beck, 2001).

The success of lift irrigation systems built by the Sadguru Foundation was due to following reasons: i) use of high quality construction materials, ii) best design by experienced civil engineers, and iii) strong community mobilization combined with capacity building by experienced social workers (Agoramoorthy, 2009). Unfortunately, government agencies while building minor irrigation infrastructures do not emphasize participatory approaches. Without the participation of communities, even highly funded development projects are doomed to fail or short-lived (Jagawat 2005). For example, in Jhabua District of Madhya Pradesh, the government has installed over 1000 lift irrigation systems during the last decade, but 70 per cent of them have failed (Choudhry *et al.*, 2002). Similarly, majority (80 per cent) of the check dams built across China's Shanbei region between 1977 and 1978 has failed due to poor construction, flawed site selection, and lack of participatory approach (Xu *et al.*, 2004).

The financing of water-related infrastructure for decades was heavily dependent on government funds. The global infrastructure financing for telecommunications, power, transport and water accounted nearly half of all government spending. But the results had been perceived as unsatisfactory

(Brisco, 1999). In the last decade, about 15 per cent of the infrastructure investment in developing countries came from the private sector. Therefore there is a need for financial partnership involving government and private corporations with NGOs as catalysts to promote sustainable development. For example, along the US and Mexico border, NGOs serve as catalyst for public-private water infrastructure (Lemos *et al.*, 2001). In Jordon, a successful public-private partnership has been reported in the management of domestic water sector using metaphors from ecology (Al-Jayyousi, 2003). The tri-sector partnership is therefore essential to minimize poverty at the grassroots across developing countries.

Agriculture Sustainability

Agricultural system is considered sustainable only when productivity is maintained over a long period by enhancing conservation of natural resources with significant profitability to guarantee financial benefit for farmers (Kessler 1994). Since agricultural production is linked to surrounding environment, consideration of all interactions between agriculture and ecosystems is a requirement for evaluating sustainability (van Wiren-Lehr 2001). India needs to double the food grain productivity by 2020 for domestic consumption. Hence it is crucial to reduce the overshoot of bio-capacity. Five major factors that determine the extent of global overshoot on bio-capacity. They include population, consumption of goods/services per person, resource use intensity, bio-productive areas, and bio-productivity per acre (WWF, 2006). An easier approach to reduce overshoot of bio-productive area is to transform India's vast drylands by building adequate water harvesting and distribution structures (Shah *et al.*, 1998, Jagawat, 2005; Agoramoorthy, 2009). The fact of agriculture being an important sector in rural development can neither be ignored nor be denied (Jewitt and Baker, 2007). Drylands must be focused to increase food productivity if India aims to succeed in sustainability, without creating negative consequences to nature.

Furthermore, human population pressure has been blamed for a series of ecological calamities in the Indian sub-continent where the density is 236 people per km^2, more than a thousand times of the rural Amazon (Agoramoorthy and Hsu, 2002). The fast growing human population pressure, water shortages and desertification may increase in the region. Nevertheless, it can be averted by fundamental change in attitude that involves mobilizing community support to integrate water resources management with agricultural development, and to focus more on the national support for sustainable agriculture initiatives (Gleick, 1999). The best way to integrate water resources management in drylands is by controlling the rainwater run-offs by constructing series of check dams in rivers (Agoramoorthy and Hsu, 2013).

India's economy is the fourth largest in the world in terms of purchasing power parity with a GDP of US$ 4.042 trillion. Despite its impressive growth, the Prime Minister had admitted that the dividends of growth failed to trickle down to the impoverished farmers (Singh, 2007). In order to eradicate poverty, India must strengthen rural development by integrating people with government, non-government and corporate support. Several NGOs have integrated the above on a smaller scale (Phansalkar and Verma, 2005; Shah *et al.*, 1998; Minj, 1999; Kashwan, 2006). A common perception in India is that NGOs complain more, but deliver less; they hinder economic progress.

Some have been implicated in misusing funds while others did excellent work, especially in water resource management, education, organic farming, and health (Anand and Anand, 2007). But they are often small, restricted to particular region to assist only limited communities and still struggling to find models that can be successfully implemented across India's diverse regions. So, it is imperative for NGOs that are involved in rural development to have transparency, accountability and responsible utilization of public funds. Besides, social responsibility should not be limited to large corporations and greater participation from small businesses and NGOs are absolutely necessary. Even small NGOs can significantly contribute for socioeconomic and political changes at the grassroots level. For example, the Sadguru Foundation's programs have given employment opportunities to over 1.6 million farmers through various farming activities.

Conclusion

India and China are among the top 10 nations most vulnerable for food shocks since they harbor large human populations. Food is the largest single component of household spending, nearly 80 per cent, in countries with low per capita incomes, compared with 15 per cent of the average family in EU or USA (Starke, 2008). Sustainable development in the agriculture sector may therefore hold the key for the future survival of humanity. The revival of the community lift irrigation cooperative model at the grassroots highlighted in this paper has the potential to improve agricultural output, minimize local food insecurity, reduce farmers' suicide, preserve water resources, and ultimately lessen rural poverty.

References

Agoramoorthy, G. 2008. Can India meet the increasing food demand by 2020? *Futures* 40:503-506.

Agoramoorthy, G. 2009. *Sustainable development: The power of water to ease poverty and enhance ecology.* Delhi: Daya publishing house.

Agoramoorthy, G., and M. J. Hsu, 2002.Threat of human-induced climate change cannot be ignored. *Current Science* 82:904-905.

Agoramoorthy, G., and M. J. Hsu, 2013. Partnership for poverty reduction in Gujarat, India: A case study of Sadguru foundation's water resource development work. *Asia Pacific Journal of Social Work and Development* 23:59-70.

Al-Jayyousi, O. R. 2003. Scenarios for public-private partnerships for water management. A case study from Jordon. *International Journal of Water Resources Development* 19:185-201.

Anand, R., and U. Anand, 2007. India needs its NGO. *Harvard International Review*, 8 January (hir.harvard.edu).

Anderson, D., and G. Lafond, 2010. Global perspective of arable soils and major soil associations. *Agricultural Soils of the Prairies* 3:1-8.

Atkins, P., and I. Bowler, 2001. *Food in society: Economy, culture and geography.* London: Arnold.

Bajpai, N., and J. D. Sachs, 2000. *India's decade of development.* Cambridge: Harvard University Press.

Beck, T. 2001. Building on poor people's capacities: The case of common property resources in India and West Africa. *World Development* 29:119-133.

Bhattarai, M., R. Barker, and A. Narayanamoorthy, 2007. Who benefits from irrigation development in India? Implications of irrigation multipliers for irrigation financing. *Irrigation and Drainage* 56:207-225.

Brisco, J. 1999. The changing face of water infrastructure financing in developing countries. *International Journal of Water resources Development* 15:301-308.

Bruges, M., and W. Smith, 2008. Participatory approaches for sustainable agriculture: a contradiction in terms? *Agriculture and Human Values* 25:13-23.

Chitale, M. A. 1994. Irrigation for poverty alleviation. *Water Resources Development* 10:383-391.

Choudhry, K., H. Shah, and H. Jagawat 2002. *A Study of Government-installed Lift Irrigation Schemes in District Jhabua, Madhya Pradesh*. Dahod, India: Sadguru Foundation.

Creswell, J. W., 1994. *Research design: Qualitative and quantitative approaches*. Los Angeles: Sage.

FAO, 2012. *Investing in agriculture for better future*. London: Earthscan.

Gaiha, R. 2000. Do anti-poverty programs reach the rural poor in India? *Oxford Development Studies* 28:71-95.

Gleick, P. H. 1999. The human right to water. *Water Policy* 1:487–503.

Government of India, 2001. *Census of India*. Delhi: Government of India press.

Greenland, D. J. 1997. The sustainability of rice farming. Wallingford: CAB International IRRI.

Hussain, I., M. Hanjra, 2004. Irrigation and poverty alleviation: Review of the empirical evidence. *Irrigation and Drainage* 53:1-15.

Isham, J. 2000. *The effect of social capital on technology adoption: Evidence from rural Tanzania*. IRIS Center Working Paper 235. Maryland: University of Maryland.

Jagawat, H. 2005. *Transforming the dry lands: The Sadguru story of western India*. Delhi: India Research Press.

Janaiah, A., M. L. Bose, A. G. Agarwal, 2000. Poverty and income distribution in rainfed and irrigated ecosystems: Village studies in Chattisgarh. *Economic and Political Weekly* 30:4664-4669.

Jarvis, W. T. 2014. *Contesting hidden waters: Conflict resolution for groundwater and aquifers*. New York: Routledge.

Jewitt, S., and K. Baker, 2007. The green revolution re-assessed: Insider perspective on agrarian change in Bulandshahr District, western Uttar Pradesh, India. *Geoforum* 38:73-89.

Jha, P. 2002. *Land reforms in India*. Delhi: Sage.

Kapila, K., and U. Kapila, 2002. *Indian agriculture in the changing environment.* Delhi: Academic Foundation.

Kashwan, P. 2006. Traditional water harvesting structure: community behind community. *Economic and Political Weekly* 18:596-598.

Kessler, J. J. 1994. Usefulness of human carrying capacity concept in assessing ecological sustainability of land-use in semi-arid regions. *Agriculture Ecosystem and Environment* 48:273–284.

Kurien, V. 2004. India's milk revolution: *Investing in rural producer organizations.* Washington, D. C.: World Bank.

Lemos, M. C., D. Austin, and R. Merideth. 2001. Varady, Public-private partnerships as catalyst for community-based water infrastructure development: the boarder water works program in Texas and New Mexico colonias. *Environment and Planning C Government and Policy* 20:281-295.

Lipton, M., and R. Longhurst, 1989. *New seeds and poor people.* London: Unwin and Hyman.

Mascarenhas, R. C. 1988. *Strategy for rural development: Dairy cooperatives in India.* Delhi, India: Sage.

Mikkelsen, B. 1995. *Methods for Development Work and Research: A Guide for Practitioners.* Delhi: Sage.

Minj, S. 19999. *Lift irrigation to lift from poverty.* Delhi: Manak Publication.

Narayan, D. 1995. *Designing community-based development.* Washington, D.C.: World Bank.

NCRB, 2014. *Accidental death and suicide in India.* Delhi: National Crime Records Bureau, Ministry of Home Affairs.

North, D. 1990. *Institutions, institutional change, and economic performance.* Cambridge: Cambridge University Press.

Ostrom, E. 1992. *Crafting institutions: Self-governing irrigation systems.* San Francisco: Institute for Contemporary Studies.

Phansalkar, S. J., and S. Verma, 2005. *Mainstreaming the margins.* Delhi: Angus and Grapher.

Pretty, J. 1995. Participatory learning for sustainable agriculture. *World Development* 23:1247-1263.

Renner, J. 2012. *Global irrigated area at record levels.* Washington, D.C.: Worldwatch Institute.

SAS Institute, 2000. *SAS/ETS software: changes and enhancements, Release 8.1.* Cary, North Carolina: SAS Institute.

Shah, M., D. Banerji, P. S. Vijayshankar, and P. Ambasta, 1998. India's drylands. *Tribal societies and development through environmental regeneration.* Delhi: Oxford University Press.

Sharma, N. D. 2007. Do you believe in a second green revolution? *Current Science* 92:1032-1033.

Shiva, V. 1992. *The violence of the green revolution: Ecological degradation and political conflict in Punjab.* Delhi: Zed Press.

Singh, M. 2007. www.hindu.com/nic/pmspeech070815.htm.

Singh, K., and V. Ballabh, 1996. *Cooperative management of natural resources.* Delhi: Sage.

Singh, S., T. Coelli, and E. Fleming, 2001. Performance of diary plants in the cooperative and private sectors in India. *Annals of Public and Cooperative Economics* 72:452-479.

Sood, J. 2014. New milky way. *Down to Earth* 1:26-35.

Starke, L. 2008. *State of the world: Innovations for a sustainable economy.* New York: WW Norton and Company.

van Wiren-Lehr, S. 2001. Sustainability in agriculture—an evaluation of principal goal-oriented concepts to close the gap between theory and practice. *Agriculture Ecosystem and Environment* 84:115–129.

Varshney, A. 2000. *Ethnic conflict and civic life: Hindus and Muslims in India.* New Haven: Yale University Press.

WWF, 2006. *Living planet report.* Gland, Switzerland: WWF International.

Xu, X. Z., H. W. Zhang, and Q. Zhang, 2004. Development of check dam systems in gullies on the Loess plateau, China. *Environment Science and Policy* 7:79-86.

Chapter 3

Community Irrigation Relieves Poverty

"A hungry man is not a free man"

— Adlai Stevenson

Introduction

Irrigation is fundamental for economic growth and poverty reduction since India leads the world in holding the maximum irrigated area (Droogers, 2002). Agriculture is the primary source of livelihood for 75 per cent of India's over 1 billion people. Majority of the Indian workforce (68 per cent) relies on farming despite India's agriculture contribution to the gross domestic product has diminished from 38 per cent in 1975 to 19 per cent in 2008 (Agoramoorthy, 2009). India's indigenous communities live in the drylands; they are often faced with low and erratic rainfall in addition to droughts. They are also economically weak since they have less access to public services. From the government point of view, a single tribal hamlet is too small to economically justify a school or health center, and the lack of purchasing power make the village business not at all viable (Janaiah *et al.*, 2000).

The worst social problem facing India now is poverty so the social work education principles, philosophies and practices have been criticized for failing to prepare the economically and socially weak communities to actively participate in poverty reduction schemes sponsored by various government agencies (Nanavatt, 1990; Barnabas *et al.*, 1996; Cox and Pawar, 2006). Several non-government organisations (NGOs) have been involved in the rural development work across rural India (Min, 1999), and some such as the Professional Assistance for Development Action, BAIF Development Research Foundation, Ralegan Siddhi Pariwar, Seva Mandir and Tarun Bharat Sangh work entirely on water resources (Kashwan, 2006; Mehta and Satpathy, 2008). Although India's agriculture sector calls for the application of social work skills, less research has been done in the area of irrigation-based social work. In this chapter, I present data on irrigation-based social work practiced by Sadguru Foundation to demonstrate how it could enhance livelihoods of the impoverished farmers across the drylands of western India.

Background

Irrigated agriculture has a long history and ancient Indian scriptures have made references to the benefits of wells, canals, and dams to society. Built during the 2nd century AD, the Grand Anicut is considered to be the oldest irrigation system in the world, which is still operational (Agoramoorthy, 2009). The historical record keeping of irrigated areas began in the 1800s with 8 million hectares (M ha) and it increased to 278 M ha by 1999 (FAO 2001). Most of the world's irrigated area has been distributed in six countries namely India (21.7 per cent), China (19.4 per cent), USA (7.9 per cent), Pakistan (6.6 per cent), Iran (2.8 per cent) and Mexico (2.4 per cent; Droogers 2002). These countries also have the highest proportions of irrigation relative to cultivated area (50.1 per cent India; 49.8 per cent China, 21.4 per cent USA; 17.2 per cent Pakistan, and 7.3 per cent Iran; Postel 1999). However, the agricultural system is considered sustainable only when productivity is maintained by profitability and sustainability (Kessler 1994; van Wiren-Lehr 2001).

Unlike the non-tribals who prefer to live in cities, tribal communities often inhabit remote wilderness region covering India's vast drylands (Jha 2002). They own land but of poor quality with no irrigation facilities. Although irrigation has been known to alleviate poverty (Chitale 1994; Brisco 1999; Bhattarai *et al.*, 2007), the gravity irrigation practiced usually along the plains and in flatlands is not at all suitable for the rugged drylands where farms are at a higher level than the rivers (25-40 m). Thus lifting water directly from rivers or from the smaller check dams built across rivers to upland farms using mechanized pumps is commonly known as 'lift irrigation' and it is considered to be the best option for the rugged terrains (Agoramoorthy and Hsu 2008). But, financial support is crucial to build lift irrigation systems and check dams and then only rainwater can be harvested, stored in check dams and used for irrigation. The global infrastructure investment for power, transport and irrigation water accounts nearly half of all the government spending. But the results in terms of benefit to people have been unsatisfactory in the past decades (Brisco, 1999).

In the recent years, India's rain-fed agriculture has been dwindling due to erratic monsoons associated with global warming. Debt-ridden farmers who lose crops due to crop failure often commit suicide and the numbers are increasing annually (NCRB, 2007). Faced with an ever growing population, India has no choice except to double the food grains productivity by 2025 to guarantee food security. The fact that agriculture plays an important sector in rural development cannot be denied (Jewitt and Baker, 2007). By 2050, humanity will demand resources at double the rate at which the Earth can generate. Therefore agriculture sustainability has become a priority in recent years (Borch, 2007; Agoramoorthy, 2009). Five conditions are crucial for an agricultural system to be sustainable and they are: the system must be ecologically sound, socially just, economically viable, humane for humanity, and adaptable to changing climate (Reijntjes *et al.*, 1992). Without the participation of farming communities in rural areas even the highly funded development projects are doomed to fail. For example, in the Jhabua District of Madhya Pradesh (India), the government had installed 1,000 lift irrigation systems of which 70 per cent ended in failure (Choudhry *et al.*, 2002). Similarly, 80 per cent of the check dams built in China failed due to poor construction and lack of community participation (Xu *et al.*, 2004).

Methodology

The objective of this study is to demonstrate how social work can be employed in India's dryland villages to develop community-irrigation to enhance agriculture and improve livelihoods of the impoverished farmers. This paper presents data on irrigation-based projects implemented by Sadguru foundation in the states of Gujarat (Dahod district) and Rajasthan (Jhalawar and Banswara districts) between 1977 and 2008. Villages located in Gujarat and Rajasthan states were visited between January 2007 and

December 2008 to record data on lift irrigation systems set up in check dams and their benefits to the dryland farmers. Information on lift irrigation operation, water source, lifting heights, total cost, irrigated area expansion and number of beneficiaries was pooled from the archives of Sadguru foundation. A total of 250 farmers were randomly selected in villages to gather data on households, sex ratio, land holdings, irrigation area, crops grown, security of food grains and immigration patterns. Major monsoon crops (*Kharif* in Hindi language) recorded during this study were maize and paddy (June-September) while post-monsoon crops (*Rabi* in Hindi language) included wheat, maize, gram, and pulses (October-February). Statistical Analysis System software was used for analyses, and all mean values are presented as ± 1 standard deviation (SAS Institute, 2000). Spearman correlations were used to test the relationships among the number of lift irrigation system, the number of farmers and the size of irrigated area. The General Lineal Model was used to test the effect of state (Gujarat/Rajasthan) and number of lift irrigation systems on the dependent variables (expansion of irrigated area).

Community Benefits of Lift Irrigation

A total of 286 lift irrigations were established in Gujarat and Rajasthan states benefiting 120,552 farmers (average lift irrigation/taluka = 19.1 ± 20.2, n=15, range 1-76; average irrigation cooperatives/taluka = 18.1 ± 19.1; Table 3). The average number of benefitted farmers was 8036.8 (± 9719.1, n=15) and the expansion of irrigated area averaged 1362.5 acres (± 1655.77, range 24-5951; Table 3).

Gujarat state had 189 lift irrigations with 181 cooperatives while Rajasthan state had 97 lift irrigations with 91 cooperatives and it shows the increasing support from farmers, government agencies and donors. The total expansion of irrigated area was 20,438 acres, and 70.8 per cent was in Gujarat state benefitting 90,774 farmers. Gujarat state's Jhalod taluka had the highest number of lift irrigation (26.6 per cent) and cooperatives (n=72) benefitting 34,722 farmers and expanding 5,951 acres of irrigated land (Figure 10). The number of lift irrigations was significantly correlated with the number of beneficiary and size of irrigated area (P<0.001, Spearman Correlation, n=15, Table 4). Furthermore, the number of lift irrigation had significant impact on the size of irrigated area ($F_{1, 12}$=207.27, p<0.001; Figure 10). Case studies of crop yield recorded in 6 villages namely Parasali,

Table 3: Community Irrigation Cooperatives Established in Villages of Gujarat and Rajasthan States of India till December 2008

State Name	Taluk Name	No. of Lift Irrigation Systems	No. of Irrigation Co-operatives	No. of Members	No. Benefitted farmers	Expansion of Irrigated Areas (Acre)
Gujarat	Dahod	42	42	3859	23316	3156
	Garbada	7	7	1300	5550	854
	Fatepura	5	4	378	1734	581
	Limkheda	32	30	2609	15126	2731
	Devgarg Bariya	16	15	643	6756	672
	Jhalod	76	72	6067	34722	5951
	Dhanpur	11	11	503	3570	518
Rajasthan	Kushalgarh	36	31	2032	10074	2475
	Bagidora	15	15	500	3852	563
	Choti sarvan	1	1	29	348	24
	Gadhi	3	2	99	714	112
	Aaspur	4	4	172	684	141
	Simalwada	11	11	346	1860	164
	Dug	21	21	1359	9672	2246
	Bhawanimandi	6	6	254	2574	250
Mean		19.1	18.1	1343.3	8036.8	1362.5
SD		20.2	19.1	1698.5	9719.1	1655.77
SEM		5.2	4.9	438.6	2509.5	427.5

SD: Standard deviation; SEM: Standard error of the mean.

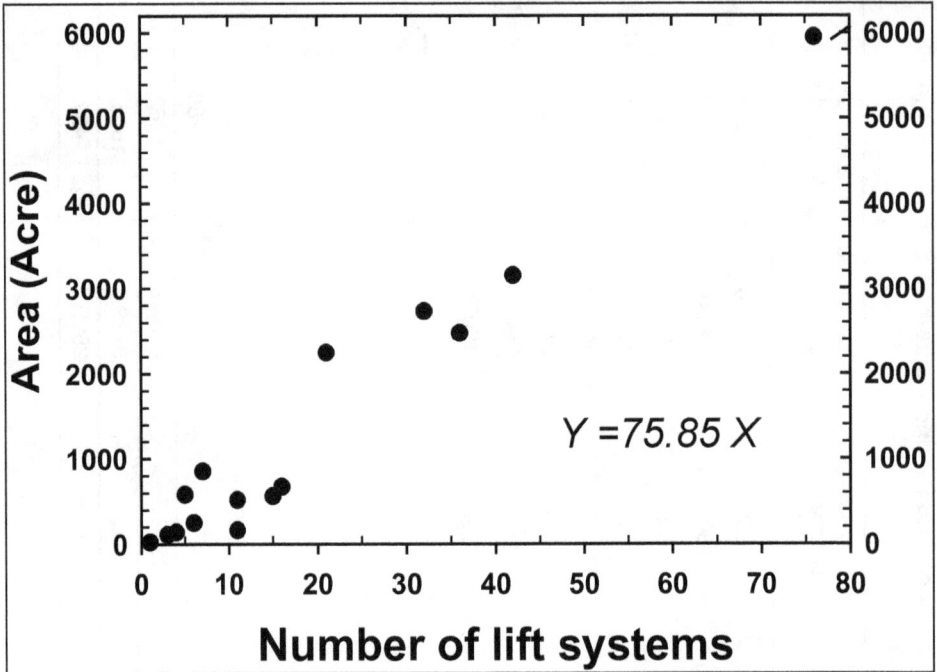

Figure 10: The Relationship of Irrigated Area Expansion (Y) and Number of Lift System (X) in Villages of Gujarat and Rajasthan States, India (regression line Y= 75.85 X).

Khejadia, Motizer, Zerjitgadh, Motadharola and Sampoi showed the highest agricultural productivity in the post-monsoon season for grams and pulses (85.1 per cent). The productivity has increased from 96.3 kg/acre (± 88.2) to 648.17 kg/acre (± 370.3). Similarly, the post-monsoon maize yield after the establishment of lift irrigation systems also increased dramatically to 79.2 per cent (from 140 kg/acre ± 198 to 671.5 kg/acre ± 449.0). Prior to the involvement of social workers, all households in the above villages had

Table 4: Spearman Correlation Coefficients Matrices for 15 Taluk-Level Irrigation Cooperatives in Gujarat and Rajasthan State

	No. of Lift Irrigation Systems	No. of Members	No. Benefitted Farmers	Expansion of Irrigated Areas
No. of lift irrigation systems	1.000	0.938	0.962	0.904
No. of members	0.938	1.000	0.975	0.982
No. benefitted farmers	0.962	0.975	1.000	0.957
Expansion of irrigated areas	0.904	0.982	0.957	1.000

All correlations were significantly different from zero. (p<0.001).

faced food shortages so farmers had to migrate to nearby towns and cities in search of work. After the social workers set up the lift irrigation systems and taught community members on how to manage the irrigation water through the village-level irrigation cooperatives, farmers attained self-sufficiency in food grain production leading to the end of seasonal migration. The lift irrigation systems also played a major role in reducing the usage of diesel engines and fuel cost. For example, a farmer had to spend one US dollar on an hourly basis previously due to the use of diesel engine. But after switching to electricity-based lift irrigation system, 50 per cent of the diesel cost was reduced.

Applied Social Work in Community Irrigation

The idea of irrigation-based social work is based on the fact that drylands receive adequate monsoon rains and if harvested, it will greatly improve agriculture in the water-scarce drylands. The Sadguru foundation has two founding directors and both have a Master Degree in social work. They lead a team of 20 social workers (with social work university education) and 30 irrigation engineers. Farmers knew this well hence they approached the foundation's social workers for help. The foundation's directors then assigned two social workers for each village who discussed with community members to know their needs for water harvesting structures. During the meetings, social workers motivated farmers to mobilize themselves to form irrigation cooperatives so that the lift irrigation infrastructure could be maintained by people. Afterwards, civil engineers accompanied by social workers conducted field surveys to fix the lift irrigation distribution chambers and water lifting points in villages. The participation and consultation of farmers were crucial during the survey because of the necessity to learn about the past history of watersheds. Afterwards, social workers had to write grant proposal to raise funds through government agencies and private donors to construct irrigation infrastructures. The foundation raised a total of US$ 1.46 million on average during 1997-2008 from government agencies (Water Resources, Tribal Development and Rural Development). The second largest funding of US$ 0.70 million came from international sponsors (Ford Foundation, Aga Khan Foundation and Norwegian Agency for Development). It was followed by corporate sponsors namely Mafatlals, Jamsetji Tata Trust, Sir Ratan Tata Trust and Sir Dorabji Tata Trust (US$ 0.58 million).

After setting up of community-level cooperatives in villages by social workers, farmers were eligible to get water from the participatory system. Each cooperative was managed by a committee of 12 members and they were elected to represent the community while two social workers serve as advisors. The committee was headed by a chairman and assisted by a secretary, who kept financial records and organized monthly meetings. The cooperative oversaw auditing, water distribution, collection of fees, electricity bill payment, staff salaries, maintaining/repairing machineries, and solving irrigation disputes. When farmers were unable to pay dues, they were allowed to pay in full before the next irrigation season with an annual interest of 20 per cent. The cooperatives saved money and profits were deposited in banks indicating financial self-sufficiency at village level. It also shows how well farmers managed their cooperatives with the support of social workers. Sadguru foundation gave technical trainings to community members to learn skills in machinery maintenance, water pricing, record keeping, accounting, public relations, conflict resolution, fund-raising, business, leadership, and natural resource management. The applied social work at village-level contributed in the empowerment of community members who became self-reliant and confident to manage irrigation projects. Social workers paid special efforts to ensure participation of women in the trainings since they involve in the decision making process of rural agriculture activities. They also continue to work with communities to manage the lift irrigation systems and village-level cooperatives.

The involvement of social workers at grassroots level is critical during crisis. For example, in Alawa village of Rajasthan (population 1000), farmers started pumping water using diesel engines from two check dams. The dams were built in Ahu River in 2007 with a storage capacity of 25 mcft. The dams irrigated 500 acres benefitting 650 people. By seeing the progress, neighboring villagers started competing for water by placing long hoses for 3-6 km to transport water. The conflict among farmers became intense in January 2008 so they approached social workers for help. Subsequently, a meeting of all parties involved including the senior authors to resolve the conflict. At the meeting, two social workers from Sadguru foundation provided solutions. Finally, all parties agreed to implement the following to resolve the standoff: (i) source for low water consuming crops, (ii) set up rules to restrict water use by farmers from adjoining villages, (iii) minimize water wastage, (iv) reduce the use of the polluting diesel engines and replace with electrified motors managed by community, and (v) discuss with engineers to increase the height of check dams to hold more water. The above solutions coincided with the recommendations of the International Water Management Institute that called for the wise use of water for agriculture. It also showed how the social workers played a crucial role in mitigating the conflict involving water resources among two villages (Molden, 2007).

Empowering Communities through Social Work

By working with farmers in villages across the dryland stets of Gujarat and Rajasthan, social workers from Sadguru foundation created several village-level cooperatives and registered under India's Cooperative Societies Act. Between 1980 and 2010, a total of 2,577 village institutions were created that include lift irrigation cooperatives, horticulture groups, women self-help groups, watershed associations and dairy cooperative with a total membership of 334,259. The creation of such cooperatives is fundamental not only to manage water infrastructures, but also to improve livelihoods of local people. Similarly, diverse social network of civic associations are known to confront poverty, resolve social disputes and provide opportunities for community development (Midgleya and Livermore 1998; Varshney 2000).

Besides, Sadguru foundation's social workers gave direct employment to farming communities while constructing irrigation infrastructures. Between 1985 and 2010, 430,719 person days with 62.7 per cent involvement of women benefitted through the employment. Moreover, farmers usually relied on migration to nearby towns in search of jobs and after the construction of check dams and lift irrigation systems, farmers had access to irrigation water and it stopped the migration. The average number of people migrating before in villages was 74.58 (± 40.36, range 18-143, n=12) with 178.83 average days (± 22.44, range 137-219, n=12). Afterwards, the number of people migrating to towns and cities reduced to 25.58 (± 13.30, range 6-48, n=12) with 57.5 average days (± 30.26, range 9-111, n=12). The reduction in migration showed gender differences as well, and for example, migration of men reduced by 61 per cent, from an average of 45.75 (±18.85, n=12) to 17.83 (22.44, n=12), while women migration reduced even higher by 73 per cent, from an average of 28.83 (±23.97, n=12) to 7.75 (7.03, n=12). This indicates that irrigation-based social work can minimize migration and maximize livelihood opportunities for farmers.

Discussion

A total of US$ 10.75 million was spent to complete all the lift irrigation structures between 1990 and 2008. In 2008 alone, the irrigated land area increased to 1575 acres from 13 completed check dams, followed by 1303.5 acres from 9 check dams completed in 1995. Most of the lift irrigation established in rivers (82.6 per cent) benefited people between 1995 and 2008 with an expansion of 10,619 acres. Over all, US$ 11.31 million was spent between 1977 and 2008 to complete the lift irrigation schemes that expanded 43,900 acres of irrigated land ultimately helping 136,000 people.

Sadguru foundation has gained a reputation in building high quality irrigation infrastructures with experienced engineers supported by professional social workers (Jagawat, 2007). Besides, participation of community members at grassroots level in villages is important, and then only rural development work will succeed (Cox and Pawar, 2006). Unfortunately, government agencies across India while building minor irrigation infrastructures such as check dams and lift irrigation systems do not emphasize participatory approach and seldom consult with social workers, and as a result, many schemes often ended in failure (Choudhry *et al.*, 2002; Jagawat 2005).

The government of the desert state of Rajasthan is more sympathetic to farmers even now by subsidizing or cancelling costly electricity bills during drought years. This approach can be extended to other dryland states across India to help farmers in crisis since they lack crop loss insurances. Another issue affecting rural farmers is the shortage of electricity. Unfortunately, India does not produce adequate electricity. It generates 129,000 MW of power annually but the demand is 200,000 MW. Farmers in Rajasthan get only 6–8 hours of electricity daily while in Madhya Pradesh only 2–4 hours. The Prime Minister of India has called the electricity shortage as a national emergency and he admitted the lack of foresight on the part of policymakers to solve this looming social crisis (Singh, 2007). It is hoped that India's bilateral nuclear agreement with USA and the growing renewable energy options involving solar and wind power may help to ease the power distress in future (Agoramoorthy, 2012).

When people collaborate to create their own social rules, opportunities for individuals and collective empowerment can emerge (Ostrom, 1992;

Singh *et al.,* 2001). Participatory approaches to manage irrigation, agriculture, forestry, mining, watershed, and fisheries have been known to improve livelihoods (Pretty, 1995; Beck, 2001). When rural farmers face economic hardships due to sudden crop failure, the final safety net of survival constitutes their friends, family members and social workers. The formation of grassroots irrigation cooperatives collectively formed by community members with the assistance and leadership of social workers serve as a form of 'social capital' with a potential to cease social and economic crisis. When farmers form diverse social networks via civic associations similar to the irrigation cooperatives highlighted in this paper, they are in a better position to solve problems related to economic loss, crop failure, poverty, social conflicts and suicide (Narayan 1995; Isham 2000; Varshney 2000). In fact, India's crop failures and bankruptcy have led to about 15,000 farmer suicides each year (NCRB, 2007). India being a signatory to the World Trade Organization, it is under pressure to open its market to globalization. So, the poor farming communities will certainly need assistance that's why the irrigation-based applied social work presented in this chapter has great potential to reduce poverty by reviving economic development at the grassroots level.

References

Agoramoorthy, G. (2009). Sustainable development: The power of water to ease poverty and enhance ecology. Delhi: Daya Publishing House.

Agoramoorthy, G. (2012). 'India should exploit renewable energy', *Nature* 481: 145.

Agoramoorthy, G. and M.J. Hsu (2008). 'Reviving India's grassroots social work for sustainable development', *International Social Work* 51: 544-555.

Barnabas, A.P., P.D. Kulkarni, M.C. Nanavatty and R.R. Singh (1996). 'The social development summit and the developing countries', *Social Development Issues* 18:79-84.

Beck, T. (2001). 'Building on poor people's capacities: The case of common property resources in India and West Africa', *World Development* 29: 119-133.

Bhattarai, M., R. Barker and A. Narayanamoorthy (2007). 'Who benefits from irrigation development in India? Implications of irrigation multipliers for irrigation financing', *Irrigation and Drainage* 56: 207-225.

Borch, K. (2007). 'Emerging technologies in favor of sustainable agriculture', *Futures* 39: 1045-1066.

Brisco, J. (1999). 'The changing face of water infrastructure financing in developing countries', *International Journal of Water Resources Development* 15: 301-308.

Chitale, M.A. (1994) 'Irrigation for poverty alleviation', *Water Resources Development* 10: 383-391.

Choudhry, K., H. Shah and H. Jagawat (2002) *A study of Government-installed Lift Irrigation Schemes in District Jhabua, Madhya Pradesh*. Dahod: Sadguru Foundation.

Cox, D. and M. Pawar (2006). *International social work*. New Delhi: Sage.

Droogers, P. (2002). *Global irrigated area mapping: overview and recommendations*. Colombo: IWMI.

FAO (2001). Food and Agriculture Organization database (www.apps.fao.org).

Government of India (2001). *Census of India 2001*. Delhi: Government of India press.

Isham, J. (2000). *The Effect of Social Capital on Technology Adoption: Evidence from Rural Tanzania*. IRIS Center Working Paper No. 235.

Jagawat, H. (2005). *Transforming the dry lands. The Sadguru story of western India*. Delhi: India Research Press.

Janaiah, A., M.L. Bose and A.G. Agarwal (2000). 'Poverty and income distribution in rainfed and irrigated ecosystems: village studies in Chattisgarh', *Economic and Political Weekly* 30: 4664-4669.

Jewitt, S. and K. Baker (2007). 'The green revolution re-assessed: Insider perspective on agrarian change in Bulandshahr District, western Uttar Pradesh, India', *Geoforum* 38: 73-89.

Jha, P. (2002). *Land reforms in India*. New Delhi: Sage.

Kashwan, P. (2006). 'Traditional water harvesting structure: community behind community', *Economic and Political Weekly* 18: 596-598.

Kessler, J.J. (1994). 'Usefulness of human carrying capacity concept in assessing ecological sustainability of land-use in semi-arid regions', *Agriculture Ecosystem and Environment* 48: 273–284.

Mehta, A.K. and T. Satpathy (2008). *Escaping poverty: the Ralegan Siddhi case*. CPRC Working Paper No. 119. Chronic Poverty Research Centre, London, UK.

Minj, S. (1999). *Lift irrigation to lift from poverty*. Delhi: Manak Publication.

NCRB (2007). *Accidental death and suicide in India*. Delhi: National Crime Records Bureau, Ministry of Home Affairs, Government of India.

Molden, D. (2007). *Water for Food. Water for Life*. London: Earthscan.

Nanavatty, M.C. (1990). 'Changing pattern of social work education in developing countries with special reference to India: Need for alternate models', *Indian Journal of Social Work* 2: 309-321.

Narayan, D. (1995). *Designing community-based development*. Washington, D.C.: World Bank.

Ostrom, E. (1992). *Crafting institutions: Self-governing irrigation systems*. San Francisco: Institute for Contemporary Studies.

Postel, S. (1999). *Pillar of sand: Can the irrigation miracle last?* New York: W.W. Norton.

Pretty, J. (1995). 'Participatory learning for sustainable agriculture', *World Development* 23: 1247-1263.

Reijntjes, C., B. Haverkort and A. Waters-Bayer (1992). *Farming for the future: An introduction to LEISA*. Basingstoke: Macmillan.

SAS Institute (2000). *SAS/ETS software: changes and enhancements. Release 8.1*. Cary: SAS Institute, North Carolina.

Singh, M. (2007). 'Power shortages of this magnitude can be a serious impediment to economic activity', *Times of India*, 30 May.

Singh, S., T. Coelli and E. Fleming (2001). 'Performance of diary plants in the cooperative and private sectors in India', *Annals of Public and Cooperative Economics* 72: 452-479.

van Wiren-Lehr, S. (2001). 'Sustainability in agriculture—an evaluation of principal goal-oriented concepts to close the gap between theory and practice', *Agriculture Ecosystem and Environment* 84: 115–129.

Varshney, A. (2000). *Ethnic conflict and civic life: Hindus and Muslims in India*. New Haven: Yale University Press.

Xu, X.Z., H.W. Zhang and Q. Zhang (2004). 'Development of check dam systems in gullies on the Loess plateau, China', Environment Science and Policy 7: 79-86.

Chapter 4

Vegetable Farms Enhance Ecology and Livelihoods

"Farming is a profession of hope"

— *Brian Brett*

Introduction

Desertification is a serious environmental issue worldwide with major implications for environment and sustainable development (D'Odorico *et al.,* 2013). The fragile drylands further degradation will impact human ecology gravely and experts have warned that 50 million people will be displaced as the result of desertification within the next decade (Haag, 2007). The drylands cover 41 per cent of our planet's surface area and they also sustain 38 per cent of the total world population (MEA, 2005). Therefore the UN has focused on the thorny issues concerning the growing desertification crisis in drylands by adopting the Convention to Combat Desertification in 1992 (Chasek, 1997). The UN also designated 2006 as the International Year of the Desert and Desertification to bring awareness on the need to promote sustainable development to decelerate ecological deterioration.

The drylands dominate India since large landmass, 564 million acres out of the total land area of 810 million acres, falls under the category of drylands (National Report to UNCCD, 2010). These enormous drylands are vulnerable for climate change. India has the second largest arable land in the world following the United States. But the Indian agriculture is dominated by small farms where 60 per cent of landholders own 17 per cent of farmlands with an average holding of 2.5 acres. In contrast, 7 per cent of medium-to-large landholders (>10 acres) own 40 per cent of farmlands (European Commission 2007). Moreover the small landholders are subsistence farmers with low investment potential so they cannot sustain farming without strong financial backing. The collapse of Doha 2006 WTO Development Round negotiations has showed the alarming trend (Pritchard, 2009).

About 68 per cent of India's workforce relies on farming in comparison to 44 per cent in China and 21 per cent in Brazil (Tripathi and Prasad, 2009). Nonetheless, the country's agriculture contribution to GDP has diminished from 38 per cent in 1975 to 14 per cent 2012 (Agoramoorthy, 2012). Women who live in rural areas are more vulnerable to poverty due to less ownership of farmlands (2 per cent) therefore access to farming and credit affects them greatly (Berger, 1989). Addressing the linkage between poverty and agriculture has become fundamental for India due to its growing population in the ecologically-sensitive drylands (Dev, 2007). For that reason, role of small rural enterprises in poverty alleviation has become extremely critical (Ellis, 2000; Dubey, 2008; Gupta, 2011).

There are not many economic opportunities in India's drylands to enhance natural resource management therefore NGOs are extending assistance to government agencies on the task of empowering women through micro-enterprises (Kraus-Harper 1998; Meyer and Nagarajan 2000; Choudhary, 2012). One such intervention is the implementation of vegetable scheme. This chapter presents data on vegetable farming by small-farm holders, especially women. It also discusses further on the economic benefits and ecological sustainability of vegetable farming in India's drylands that are often susceptible for environmental degradation.

Methodology

This case study on vegetable cultivation was carried out in Dahod District (area 3,642 km²) of Gujarat State, India. Between January 2012 and August 2013, data on the impact of vegetable farming in Dahod district were collected from 100 women farmers in 27 villages that spread across six talukas (similar to county) namely Dhanpur (n=23), Limkheda (n=21), Jhalod (n=20), Zalod (n=20), Devgadh Bariya (n=10) and Bariya (n=6).

Between 2000 and 2004, 31 women farmers joined Sadguru foundation to get trainings on vegetable cultivation while the rest (n=69) joined between 2005 and 2010. The largest number of women farmers (n=42) joined in 2010 after seeing other farmers' income gain from vegetable farms. During visits, data on name of village, household information, poverty level, family size, and economic benefits derived from vegetable farms were recorded by adopting the methods described in Barnet *et al.* (1995) and McNeely and Scherr (2001).

All statistical analyses were conducted using Statistical Analysis System software (SAS Institute, 2000). All mean values are presented as ± 1 standard deviation (SD). Local currency (India rupee) was converted to US dollars (conversion rate: 1 US$=50 Rs). Paired t-tests were used to compare the average amount of monthly income, bank balance and yearly expenses for medical treatment of households before and after joining the vegetable scheme. Paired t-tests were also used to compare the total number of tools per family, frequency of daily meals consumed and annual hospital visits of households.

Results

The average household size was 7.1 ± 2.7 (adults 4.0 ±1.8, children 3.2 ± 1.5). Only 39.3 per cent of the people were literate (±24.3 per cent). All households in general have their own farms with an average size of 3.0 ±1.3 acres. Prior to the cultivation of vegetable, the farmers lived mainly in thatched roofs (86 per cent); their houses were made of mud, bamboos, straw and leave. Only 14 per cent of them had better housing made of bricks, and cement. The average annual income of households was US$ 108.8 ± 80.37. Only five had bank savings (US$ 260 ± 162.48, n=5, range US$ 100-500) and 13 had crop loans (US$ 169.23 ± 82.12, n=13, range US$ 100-400). The average hospital annual visits by farming households was 7.1 ± 2.9 times, with medical bills reaching up to US$ 136.70 (± 70.98, n=100). Most households in average had only 2.0 meals (±0.30) per day.

Most of the households (88 per cent) had some farming tools such as plough, spade, axe, pickaxe, design plough, cultivator and spade axe before joining the vegetable scheme. Some 43 families owned a single type of tool while 12 did not have any tools at all. However, 30 families had 5 to 9 tools. The most common tool was plough (31.82 per cent), followed by spade (22.22 per cent), axe (18.18 per cent), pickaxe (12.12 per cent), design plough (6.57 per cent), cultivator (4.55 per cent) and spade axe (4.55 per cent).

None of the farmers had tractors. Only 7 families owned vehicles while 4 families had motorbikes. The major water source was groundwater wells in villages and 42 families relied mainly on wells while 26 families entirely dependent on rivers. Nearly all families grow maize (97 per cent), but 25 per cent families only grow maize and rest families grow rice (42 families) and/or wheat (28 families) in the same time. Few farmers also grew pigeon pea, gram, eggplant and Chili.

After joining the vegetable scheme, the activities of farmers were not restricted to growing traditional crops and economically-valuable vegetables, spices and flowers were involved. Over one third included a few types of gourds (36.4 per cent) such as bitter, bottle or smooth gourds. Rest of the types included seasonal vegetables (15.79 per cent), beans (10.53 per cent), orchard (9.31 per cent), okra (8.91 per cent), tomato (4.45 per cent) and other vegetables such as pointed gourd, mango and eggplant. Garlic was the most popular spices (31.7 per cent), followed by ginger (28.5 per cent), onion (19.9 per cent) and turmeric (19.9 per cent). A total of 23 families involved in floriculture with 8 families permanently and 8 seasonally; they grew flowers such as marigold, rose and Florist's daisy or locally called *sevanti* (*Chrysanthemum morifolium*).

Interestingly the farmers' illiteracy rate reduced after they joined the scheme and persons belong to literate category increased to 57.5 per cent. Similarly, the percentage of permanent houses made up of bricks, cement and concrete has increased to 47 per cent while the traditional mud/thatched roof huts reduced. The average monthly household income increased to US$ 285.46 ± 142.36, with an average bank balance of US$ 363.66 ± 445.44.

Both the monthly income and bank balance were significantly increased after the farmers started to grow vegetables (paired t-test, $P<0.05$; Figure 1). Most families had bank savings (US$ 481.32 ± 457.06, n=76, range US$ 20-2000) after the vegetable cultivation while 24 families had even bought personal/life insurance policies (US$ 100-20,000). Besides, four families bought crop insurance (range US$ 100-400), which is rare since farmers usually get loans from local loan sharks. Most families on average had 2.41 meals (±0.49) per day, which was also slightly increased when compared to earlier ($P>0.05$). Nonetheless, the health conditions of households did not change significantly. The average hospital annual visits was 6.8 ± 3.3

times, with medical bill reaching up to US$ 120.30 (± 69.74, n=100); both the expenses were only reduced slightly compared to earlier figures (Figures 11 and 12; paired t test, NS, P>0.05).

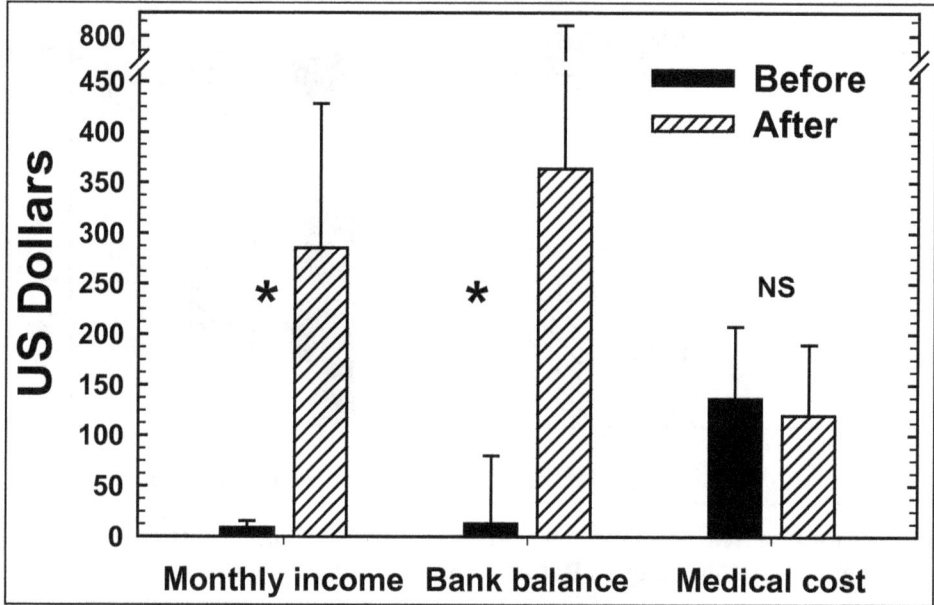

Figure 11.: Average (± SD) Amount of Monthly Income, Bank Balance and Yearly Expenses for Medical Treatment of 100 Families Before and After Joining Sadguru.

*: Paired t test, the difference significantly differed from zero at p=0.05 level. NS: none significant differences (P>0.05); Conversion rate 1 US$ =50 India Rupees.

The average total number of tools per family was also increased significantly from 2.8 ± 2.6 to 6.17 ± 3.68 (P<0.05) after the women joined the vegetable cultivations scheme. Even before joining the scheme, most families (88 per cent) had some types of tools. About 30 of them had 5-9 tools while 43 only owned one type of tools. About 12 families did not have any tool. The most common tool before was plough (31.82 per cent), followed by spade (22.22 per cent), axe (18.18 per cent), pickaxe (12.12 per cent), design plough (6.57 per cent), cultivator (4.55 per cent) and spade axe (4.55 per cent). After the farmers earned considerable profit from vegetables, plough remained the most popular tool (31.90 per cent), followed by axe, pickaxe and spade axe (22.70 per cent). The number of households owning tractors increased to six while one household bought a van and 28 bought motorcycles.

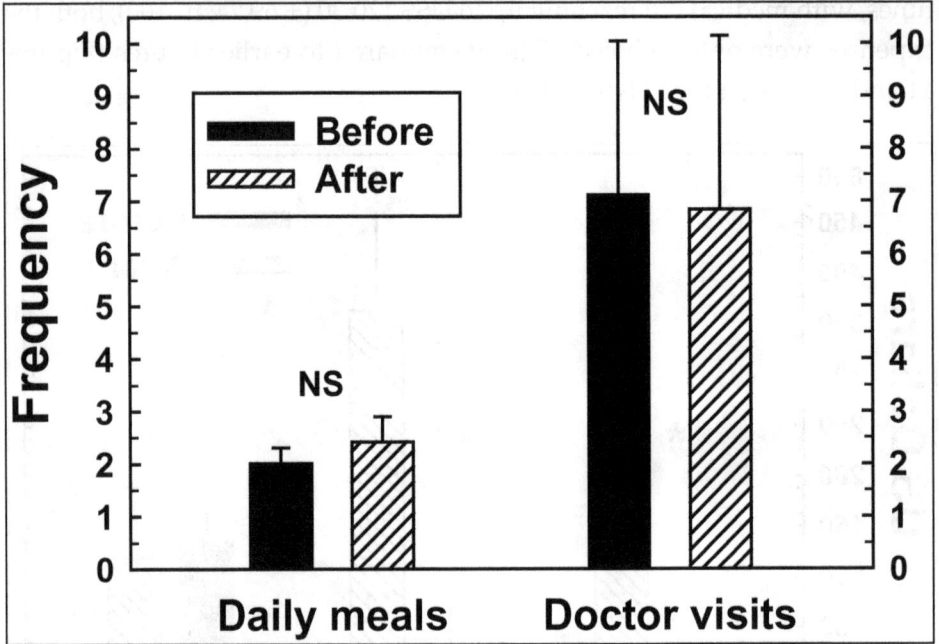

Figure 12: The Average (± SD) Frequency of Daily Meals and Hospital Visits for 100 Families Before and After Joining Sadguru.

NS: Paired t test, the difference was not significantly differed from zero at p=0.05 level.

The village vegetable farms located near the house included a compost heap with some plot. Prior to the scheme, 42 households (women and men) migrated to nearby towns for work, but afterwards, none migrated, and all remained in their farms to grow vegetables. The vegetable cultivation was done by mainly by women. Men supported them by allowing them to spend time to cultivate vegetables in the backyard. The vegetable farming was sustainable since the farmers used farmyard manure that includes animal dung, urine, straw, green leaves and other organic waste from farms and homes. Furthermore, they used organic manure generated from household biogas plants that provided eco-friendly natural fertilizer for crops and clean energy for cooking (Agoramoorthy and Hsu, 2008a). The farmers used vermin-compost that contained various microbial organisms (600 million/gram) supplying all nutrients to soil. The earthworms in vermin-compost assisted the soil to neutralize acidity, reduced salinity and soil erosion. Bu using the homemade organic manure, the farmers was able to cut the cost of fertilizers. They had to use on average US$ 200 to 400 yearly for pesticide and fertilizer earlier to grow maize and other staple crops.

Livelihoods and Socio-economic Benefits

After successful vegetable cultivation, the tribal women have gained hands-on experience in the production, marketing, and selling of vegetable; they also gained expertise in decision-making on farm activities. Food security is commonly based on three basic concepts of availability, accessibility, and utilization. But they are often hierarchical due to the necessary availability of food but not enough to guarantee accessibility, which tend to minimize effective utilization (Webb *et al.*, 2006). The monetary gain after cultivating vegetables improved livelihoods and the tribal households became economically self-sufficient.

In recent decades, the developing world has witnessed a shift in the way of food production and consumption due to the evolving choices of consumers' taste catalyzed by globalization (Dolan and Humphrey 2000; Deshingkar, 2006). Although India has doubled its per capita income in recent decades with substantial increase in population growth, the country has not witnessed a radical shift in the way in which fresh vegetables are produced. Nonetheless, statistics shows that the annual per capita consumption of fruits and vegetables between 1990 and 2000 have grown over 2 per cent when compared to cereals (DSIR, 2005). This positives trend seems to be suitable for rural farmers of drylands. This study shows that the empowerment of tribal women farmers at grassroots through exposure for agriculture technology is crucial for poverty reduction and environment protection.

Ecological Sustainability

India ranks second in the world in vegetable production, following China with 87.5 million tons, sharing 14 per cent of the world output from a mere 2 per cent of cropped area. By 2020, India will need 250 million tons of vegetable production due to increasing human population with the reduction in cultivable land (Economic Times, 2008). So, India needs to opt for the vertical expansion by increasing the productivity per unit area with less dependency on land, water, and chemicals. Vegetable cultivation in India's drylands shown here offers a sustainable farming system. However, research and development combined with financial backing from government and corporate sector is crucial if the country is serious to achieve the target to produce 250 million tons of vegetables sustainably by 2020, without depending on chemical fertilizers and pesticides.

India leads Asia in the manufacturing of pesticides (90,000 tons) and it is one of the few remaining countries engaged in the chlorinated pesticides such as DDT, HCH and PCP (Abhilash and Singh 2008). India is also the third largest producer and consumer of fertilizer in the world, after China and USA (Galloway *et al.*, 2008). The national average consumption of fertilizer has been 96.4 kg/ha with wide variation across state line (197 kg/ha in Punjab, 164 kg/ha in Haryana, and <10 kg/ha in smaller states such as Arunachal Pradesh, Nagaland, and Sikkim). However, the cost of fertilizer has been increasing in recent years hence 11 of the top 25 fertilizer-consuming countries have subsidized. China for example has paid their farmers US$ 3.7 billion in support of fertilizer, while India paid US$ 5.3 billion in farm subsidies during 2008 (Ramanjaneyulu, 2012).

So the question is: does India need to heavily dependant on chemical fertilizers? In fact, India possesses the world's largest livestock population of 485 million; it harbors a total of 57 per cent of the world's buffalo and 16 per cent of the cattle population. It ranks first in the number of cattle and buffalo, second in goat, and third in sheep and camel (Ramdas and Ghothe, 2006). Even if two-thirds of the dung is used to generate biogas from households and farms, it will yield 250 million m^3 of biogas daily. It will also yield over half billion tons of organic manure annually, equivalent to 2.90, 2.75 and 1.89 million tons of nitrate, phosphorus pentoxide and potassium oxide, respectively (Ravindranath *et al.*, 2000; Ramesh *et al.*, 2005; Agoramoorthy and Hsu, 2008b). This will supply all the required fuel for farmers and farmyard nutrients for agriculture and thus it has the potential to ease India's agriculture addiction to toxic chemicals.

The government of India has admitted that the postharvest loss of fruits and vegetables produced in rural areas leads up to the loss of US$ 30.4 billion annually due to weak marketing, poor transport/roads, inadequate power, insufficient cold storage facilities, and lack of infrastructure (Times of India, 2013). By 2025, India needs to produce 315 million tons of grains to feed its people; it has produced an average of 200 million tons during the last 10 years (Agoramoorthy, 2012). As rural women are actively involved in agricultural activities from sowing to harvesting, it's time for India to think about making future agriculture technology more women-friendly.

Unlike water-intense crops such as sugarcane, rice and wheat, vegetable crops had neither consumed excessive water nor damaged environment. Farmers had to pay electricity bill for pumping water from wells so they decreased water wastage. Besides, they used drip irrigation that reduced excess water usage and wastage. The Planning Commission of India (2007) has admitted in its tenth plan document that the water use efficiency in most canal irrigation systems connected to large dams as low (30–40 per cent, against an ideal value of 60 per cent) due to wastage, silting, weed growth, broken structures and poor maintenance (Agoramoorthy and Hsu, 2008b).

Conclusion

India's food security is heavily dependent on irrigation, yet large areas of irrigated land are threatened by increasing salinity, expanding desertification and decreasing surface and ground water. Therefore sustainable production and consumption is absolutely essential to safeguard local ecology and ecosystems. The small-scale vegetable cultivation has undoubtedly empowered tribal women to enhance their livelihoods, economy and local ecology therefore it has the potential to contribute immensely for the sustainable development of the ecologically-sensitive drylands. So this should be encouraged by both the governments and corporate sectors since it has the potential not only to enhance India's future food security but also to implement the sustainable production and consumption mandated by the Johannesburg Plan of Implementation agreed at the World Summit on Sustainable Development (2002). The plan mandates socio-economic development at the grassroots in villages within the carrying capacity of local ecosystems to decelerate climate change induced by carbon emission.

References

Abhilash, P.C., and Singh N. 2008. Pesticide use and application. An Indian scenario. *J. Hazard Mater.* 165: 1-12.

Agoramoorthy, G. 2012. *Dryland agriculture: Case studies from India.* Lambert academic publishing, Germany.

Agoramoorthy, G., and Hsu, M.J. 2006. Do animals suffer caste prejudice in Hinduism? *Social Compass* 53: 244-245.

Agoramoorthy, G. and Hsu, M.J. 2008a. Biogas plants ease ecological stress in India's remote villages. *Human Ecol.* 36: 435-441.

Agoramoorthy, G. and Hsu, M.J. 2008b. Small size, big potential: Check dams for sustainable development. *Environment: Sci. Pol. Sustain. Develop.* 50: 22-35.

Berger, M. 1989. Giving credit to women: The strengths and limitations of credit as a tool for alleviating poverty. *World Develop.* 17: 1017–1032.

Boserup, E. 1989. *Woman's role in economic development.* Earthscan publications Ltd., London.

Chasek, P.S. 1997. The convention to combat desertification: Lessons learned for sustainable development. *J. Environ. Develop.* 6: 147-169.

Choudhary, S.K. 2012. *NGOs, education and social capital: A micro study of tribals.* Satyam Books, New Delhi.

Deshingkar, P. 2006. *Internal migration, poverty and development in Asia.* ODI Briefing Paper 11. Overseas Development Institute, London.

Dolan, C.S. 2007. Market affections: Moral encounters with Kenyan fair-trade flowers. *Ethnos* 72: 239-261.

DSIR. 2005. *Fruits and vegetable sector: Overview.* Department of Scientific and Industrial Research report, Delhi.

Dubey, P. 2008. Investment in small scale forestry enterprises: Opportunities and constraints for India. *J. Busin. Persp.* 12: 39-51.

Economic Times. 2008. 72 percent of India's fruit, vegetable produce goes waste. *Economic Times*, 12 May.

Ellis, F. 2000. *Rural livelihoods and diversity in developing countries.* Oxford, Oxford University Press.

European Commission. 2007. India's role in world agriculture (ec.europa. eu/agriculture/publi/map/03_07.pdf).

Galloway, J., Raghuram, N., and Abrol Y.P. 2008. Transformation of the nitrogen cycle: recent trends, questions and potential solutions. *Science* 320: 889–892.

Government of India. 2001. *Census of India 2001.* Government press, Delhi.

Gupta, S. 2011. "Paschim Banga Kheria Sabar Kalyan Samiti", West Bengal, India: Case study of an NGO's role in poverty alleviation. *Int. J. Rural. Manag.* 7: 149-158.

Haag, A.L. 2007. Is this what the world's coming to? *Nat. Rep. Clim. Change* 5: 75-78.

Jagawat, H. 2005. *Transforming the dry lands. The Sadguru story of western India.* India research press, Delhi.

Kraus-Harper, U. 1998. *From despondency to ambitions: Women's changing perceptions of self-employment: cases from India and other developing countries.* Ashgate, Hants, UK.

Dev, S.M. 2007. *Inclusive growth in India: Agriculture, poverty and human development.* Oxford university press, Oxford.

McNeely, J.A., and Scherr, S.J. 2001. *Common ground, common future. How ecoagriculture can help feed the world and save wild biodiversity?* IUCN, Gland.

MEA. 2005. *Millennium ecosystem assessment- Ecosystems and human well-being: Desertification synthesis.* World Resources Institute, Washington, DC.

Meyer, R.L., and Nagarajan, G. 2000. *Rural financial markets in Asia: Paradigms, policies and performance.* Oxford University Press, Oxford.

National Report to UNCCD. 2010. *Elucidation of the 4th national report to UNCCD.* Ministry of Environment and Forests, Government of India.

D'Odorico, P., Bhattachan, A., Davis, K.F., Ravi, S., and Runyan C.W. 2013. Global desertification: Drivers and feedbacks. *Adv. Water. Resour.* 51: 326-344.

Phansalkar, S., and Verma, S. 2005. *Mainstreaming the margins: Water-centric livelihood strategies for revitalizing tribal agriculture in Central India.* Angus and Grapher, Delhi.

Planning Commission of India. 2007. *Irrigation, flood control and command area.* The planning commission report, Delhi.

Pritchard, B. 2009. The long hangover from the second food regime: a world-historical interpretation of the collapse of the WTO Doha round. *Agri. Human Values* 26: 297-307.

Ramdas, S.R., and Ghotge, N.S. 2006. India's livestock economy (www.india-seminar.com).

Ramanjaneyulu, G.V. 2012. Adapting smallholder agriculture to climate change. *IDS Bull.* 43: 113–121.

Ramesh, P., Singh, M., and Rao, A.S. 2005. Organic farming: Its relevance to the Indian context, *Curr. Sci.* 88:561-568.

Ravindranath NH, Rao UK, Natarajan B, and Monga P. 2000. *Renewable energy and environment: a policy analysis for India.* Tata McGraw Hill, Delhi.

SAS Institute. 2000. *SAS/ETS software: Changes and enhancements.* SAS Institute, Cary.

Times of India. 2013. Rs 44,000 crore worth of food goes water every year. Times of India 23 August (articles.timesofindia.indiatimes.com/2013-08-23/india).

Tripathi, A., and Prasad, A.R. 2009. *Agriculture development in India since independence: A study on progress, performance and determinants.* The Berkley electronics press.

Webb, P., Coates, J., Frongillo, E.A., Rogers, B.L., Swindale, A., and Bilinsky, P. 2006. Measuring household food insecurity: why it's so important and yet so difficult to do. *J. Nutri.* 136: 1404S–1408S.

World Summit on Sustainable Development. 2002. *The Johannesburg Plan of Implementation.* UN press, New York.

Index

www.ingramcontent.com/pod-product-compliance
Lightning Source LLC
Chambersburg PA
CBHW070802300326
41914CB00052B/598